The Miracle of Apple Cider Vinegar

METRO BOOKS
New York

An Imprint of Sterling Publishing
387 Park Avenue South
New York, NY 10016

ISBN: 978-1-4351-4228-2

For information about custom editions, special sales, and
premium and corporate purchases, please contact Sterling Special
Sales at 800-805-5489 or specialsales@sterlingpublishing.com.

Designed and typeset by Paul Saunders
Manufactured in China

2 4 6 8 10 9 7 5 3 1

www.sterlingpublishing.com

Practical Tips for Health, Home, and Beauty

The Miracle of

Apple Cider

Vinegar

Dr. Penny Stanway

METRO BOOKS
New York

About the author

Dr. Penny Stanway practiced for several years as a doctor and as a child-health consultant before becoming increasingly fascinated in researching and writing about a healthy diet and other natural approaches to health and well-being. She is an accomplished cook who loves eating and very much enjoys being creative in the kitchen and sharing food with others. Penny has written more than 20 books on health, food, and the connections between the two. She lives with her husband in a houseboat on the Thames in London and often visits the southwest of Ireland. Her leisure pursuits include painting, swimming, and being with her family and friends.

Acknowledgments

Thank you to my sister, Jenny Hare, for sharing my passion for food and cooking; to my husband, Andrew, for his unstinting enthusiasm in discussing apples and apple cider vinegar; and to my agent, Doreen Montgomery, for her endless encouragement and support.

In memory of John Rench, my much-loved father,
who particularly enjoyed vinegar on cockles and soft herring roe.

Contents

Introduction

Apples are the most popular fruit in the world, and I certainly love eating them! I also enjoy drinking apple juice and hard cider, and cooking with apple cider vinegar, so writing this book has been an ongoing source of fascination to me. Creating and testing the recipes has been a pleasure, and I remain amazed and intrigued by the myriad ways in which apples, apple juice, hard cider and apple cider vinegar can benefit our health.

I grew up in a nursing home run by my parents. My mother, a nursing sister, used to say, "An apple a day keeps the doctor away," and there were always apples for us to eat. Now, after my many years of experience of working as a doctor and of researching health matters as a medical writer, I am even more certain that apples and apple cider vinegar are able to promote and protect our health and general well-being. What's more, apple cider vinegar is a cheap and efficient source of help in the home.

Archaeologists believe that apple trees originated around the Caspian Sea and the Black Sea, and that people ate apples as far back as 6500 BC. Apple cultivation spread to Europe, and in the 16th century King Henry VIII instructed his fruiterer to search the world for the best varieties, so he could set up orchards in England. Apple cultivation spread to the US, Australia, New Zealand, South Africa and South America. Nowadays, more than 7,500 varieties of apple are grown worldwide, and in many countries either home-grown or imported apples are available year-round.

Historians are unsure when people first started making hard cider, but they know it was a common drink in Britain in the first century BC and think it was probably available around the Mediterranean in the 1st century AD. In contrast, archaeologists say that apple cider vinegar has been used for much longer, since traces of it have been found in Egyptian urns dating from 3000 BC.

Apples, apple juice, cider, hard cider and apple cider vinegar have a long history and are much loved around the world. In *The Miracle of Apple Cider Vinegar* I shall explain their surprising properties and many valuable uses, and I hope you'll agree that it is a really good idea to use apple cider vinegar around the home and when cooking, and to eat at least one apple every day.

A short note:

- What Americans call "cider" or "sweet cider" is known as "apple juice" in many countries, including the UK.
- What Americans call "hard cider" is called simply "cider" in many countries, including the UK.

1

What's in Apples, Apple Juice, Hard Cider & Apple Cider Vinegar

Apples

When you eat an apple, you aren't just getting a sugary source of 50 calories with some fiber and vitamin C. You are consuming a veritable treasure chest of health-promoting substances, some of which occur in such richness in only a few other foods. For example, apples are one of the most plentiful sources of pectin fiber and phenolic compounds such as quercetin. This explains why apples are the fruits most consistently associated with a reduced risk of certain chronic diseases, including heart disease, cancer and diabetes, and why consuming apples, apple juice and apple cider vinegar really can make a difference to our health and well-being.

Major nutrients

Apples are rich in fiber and sugars – including fructose, sucrose, glucose, oligosaccharides and inulin (not insulin!). An unripe apple's carbohydrate is mainly starch, but ripening converts most of this to sugars. Apples also contain useful amounts of vitamin C and potassium. A medium-size (4oz or 100g) apple, for example, supplies 8 percent of the vitamin C and 10 percent of the potassium the average adult needs each day.

Certainly apples contain extremely little fat and protein and only very small amounts of beta-carotene, B vitamins, calcium, iron, magnesium, phosphorus and zinc. But even these small amounts contribute to our necessary daily intake. The tiny amounts of trace elements such as boron and chromium are useful too. Boron, for example, is critical to the way the body uses calcium and may help prevent the bone loss of osteoporosis. It also affects the release, use and lifespan of steroid hormones.

DID YOU KNOW?

One raw, ripe apple provides as much dietary fiber as a typical serving of bran breakfast cereal.

Fiber

Dietary fiber, once called roughage, is now officially known as non-starch polysaccharides. I'll call it fiber, though it often isn't fibrous!

Not only do apples contain the non-carbohydrate fiber lignin, cellulose and hemi-cellulose and but they are also a rich source of pectin. Indeed, pectin forms 70 percent of apple fiber, and apples are richer in pectin than are any other fruits. Cellulose strengthens plant cell walls; pectin helps hold these together. Both benefit health in different ways. As water-insoluble types of fiber, cellulose and lignin passively absorb water in the digestive tract but pass through unchanged. As a water-soluble type of fiber, pectin dissolves in water in the digestive tract to form a gel. About 90 percent of the pectin in this gel is fermented by the "good" bacteria in the large intestine, releasing valuable short-chain fatty acids such as butyric acid, and aiding the absorption of calcium and other minerals.

Unfortunately, many people don't get enough fiber in their diet. In the US, for example, the average intake is only half what it should be.

One medium apple provides more than 10 percent of our daily recommended fiber intake. More than half the health-promoting fiber is in the peel and core, so it's best to eat an apple unpeeled – core and all. The perfect apple is firm, crisp and not overripe. The riper an apple becomes, the softer and mealier it is and the less pectin it contains. "Cooking" apples are green and tart even when ripe, and are good for stewing, baking and apple pies, because heat makes their pectin soft and mushy. The peel of both cooking and dessert apples, though, remains largely intact on cooking, because its cell walls are particularly rich in insoluble fiber.

Antioxidants

Antioxidants are substances that help prevent uncontrolled oxidation. Oxidation is a chemical reaction caused by free radicals (overactive oxygen particles); it is involved in many of the body's normal metabolic processes. Certain circumstances, such as a poor diet, infection, smoking and stress, increase the production of free radicals, which can lead to a chain reaction of oxidation. This uncontrolled oxidation can encourage problems such as inflammation, infection, cancer, premature aging, heart attacks and strokes.

Apples are a good source of antioxidants. Much of an apple's antioxidant content is in its peel, which is another reason to eat apples unpeeled. Apple antioxidants include vitamin C and various phenolic compounds (such as quercetin and phlorizin), plus tiny amounts of selenium and zinc. An apple contains one-and-a-half times more antioxidant capacity than 75g of blueberries, more than twice that of a cup of tea, three times that of an orange and almost eight times that of a banana.

Scientists from Cornell University in New York have found that a 4oz (100g) apple is as powerful an antioxidant as 1,500mg of vitamin C. So, given that such an apple contains only about 5.7mg of vitamin C, most of its antioxidant activity must come from other compounds. An apple's antioxidant power is greatest if eaten whole and raw soon after picking and otherwise stored in a cold dark place such as a fridge.

DID YOU KNOW?

Apples are the major source of dietary phenolic compounds in many parts of the world, including the US and Europe.

Phenolic compounds

This is an umbrella term for a group of apple phytochemicals that includes flavonoids, various acids, tannins and lignins (a non-carbohydrate type of fiber). These are all derived from phenolic acid and are sometimes called polyphenols or phenolics. Their level in apples varies from year to year, variety to variety, tree to tree, and region to region. Apple peel contains up to six times more phytochemicals than apple flesh so once again, to gain the maximum benefit always eat your apple with its peel still on. Cider apples have higher levels of phenolic compounds than do dessert and cooking apples.

Apples are particularly rich in flavonoids, the most abundant of which are quercetin and other flavonols. Quercetin is known to have antioxidant, antihistamine and anti-inflammatory actions. It is found almost only in peel, and red apples contain more than do green or yellow ones. Quercetin is relatively stable when apples are cooked.

Other apple flavonoids include proanthocyanidins (such as the plant pigment catechin, which contributes color to peel and astringency and bitterness to an apple's flavor); phlorizin; and the plant hormone genistein (*see* page 16).

An apple's phenolic acids include chlorogenic acid, p-coumaric and quinic acids. Tannins help account for any astringent flavor, and the oxidation of tannins in cut apples is responsible for their subsequent brown discoloration.

Plant hormones (phytosterols)

Apples contain very small amounts of the flavonoid genistein, which is a phyto-estrogen (plant estrogen) and an antioxidant. The amount in an apple is tiny, but nevertheless contributes to an individual's overall phyto-estrogen intake. By latching onto estrogen receptors on cells, phyto-estrogens have an estrogen-balancing action. If a woman is making unusually large amounts of her own, stronger estrogen, the occupation of cell receptors by weaker phyto-estrogens prevents her estrogen encouraging problems associated with estrogen dominance (such as heavy periods). If she isn't making enough estrogen, the occupation of cell receptors by phyto-estrogens provides some small estrogenic stimulation which can aid conditions associated with estrogen deficiency (such as hot flushes).

Organic acids

These include malic and tartaric acids. The amount of malic acid in each variety helps determine its tartness. The more malic acid and the less sugar an apple contains, the stronger its flavor and the greater its likelihood of retaining flavor when cooked. Unripe apples and cooking apples contain more malic acid than do other apples.

Wax and pesticides

Apple skin is coated in a natural protective wax which is produced by the apple itself to preserve its high water content. Washing apples with water removes about half of this, so many commercial growers coat their produce with an officially approved edible wax (such as carnauba or shellac) to make them shiny and help prevent desiccation and rotting after they have been picked and put into storage.

Apples may also have pesticide residues on their peel. The levels are monitored carefully and should be safe. However, you may prefer to give an apple a quick scrub with a vegetable brush and water (not detergent) before eating or cooking it to remove waxes and pesticides from the skin, or you can try to buy organic.

Pips

Apple pips, or apple seeds, taste slightly bitter and almond-like, due to a small amount of a substance called amygdalin. This can be converted to poisonous cyanide by an enzyme that is present in a few of the body's cells, though abundant in cancer cells. Eating a few apples complete with pips each day is highly unlikely to pose a problem and it might (though this is unproven) act against early cancer cells.

Apple scent and flavor

An apple's scent and flavor result from its particular blend of about 250 volatile compounds, which include various esters (such as ethyl-methyl-butyrate, isobutyl acetate), alcohols, aldehydes and essential oils. The aroma and flavor of some unusual varieties resemble those of melon, strawberry, raspberry, peach, lemon, fennel, cinnamon, allspice, banana or pineapple.

Apple color

This color comes from traces of chlorophyll (a green pigment), carotenoids (yellow and orange pigments) and proanthocyanidins (pigments of various hues). Red-skinned apples tend to be sweet, green ones may be tart or sweet.

Apple Juice

Apple juice is also called "cider" in the US and parts of Canada. Apple juice diluted with water, or sweetened with added sugar, must be labeled as an "apple drink" or "apple-juice beverage." Juicing apples requires mechanical pressing, which produces cloudy juice.

Other possible processes include:

- Filtering of particulate material, including cellulose, pectins and proteins. This produces clear juice.
- Pasteurization, to prevent enzymes turning sugars to alcohols, and to protect against the growth of molds and bacteria. This gives the juice a shelf life of up to two years. Unpasteurized juice should ideally be consumed as soon as possible.

The less processing that is involved in the making of juice, the smaller the loss of nutrients and other valuable phytochemicals. Cloudy (unfiltered) juice is slightly richer in fiber than is clear (filtered) juice. It is also much richer in valuable phenolic compounds, containing, for example, a much higher amount of proanthocyanidins and other antioxidants. And while cloudy juice contains around half the amount of phenolic compounds of apples themselves, clear juice contains only up to a third or so.

To make your own apple juice, *see* pages 166–170.

> **DID YOU KNOW?**
>
> One glass of pure apple juice counts as one serving of fruit in the recommended guidelines for a healthy diet.

Hard Cider

Hard cider is an alcoholic beverage made by the fermentation of apple juice. In the US, any fermented apple juice containing more than 0.5 percent ABV (alcohol by volume, meaning milliliters of alcohol per 100 milliliters of liquid) is called "hard cider." In the UK, 1.2 percent ABV cider is designated "low-alcohol" cider. More often, the alcohol content of hard cider varies from less than 3 percent ABV (for example, French *cidre doux*) to 8.5 percent ABV or more (for example, in traditional English hard ciders).

The flavor of hard ciders differs according to the apples or blends of apples used. Their color varies from very pale gold ("white cider") to rich golden brown. The trend today is to make hard cider from a single crop of a single apple variety. The flavor of such a hard cider reflects the particular blend of volatile phyto-chemicals in that variety of apple; the cider can also be sold as being of a defined "vintage." You can use any apples to make hard cider; however, many hard cider-makers like to include hard cider apples in their chosen blend. This is because these contain higher levels of tannins and more malic acid and other organic acids than do sweet "dessert" apples, and give hard cider a characteristic "bite."

Hard cider apples are grouped in four categories according to their flavor components:

Bittersweets are high in sugar, which raises the hard cider's alcohol content. They are also relatively high in tannins, so the cider is quite bitter.

Bittersharps are high in tannins and fruit acids (for example, malic acid), so the hard cider is relatively bitter and sharp.

Sweets are high in sugar, which raises the hard cider's alcohol. They are low in tannins and fruit acid, so the hard cider has little bitterness or sharpness.

Sharps are high in acidity, which adds sharpness to their hard cider. They are low in sugar and tannins, so the hard cider is a little less alcoholic or bitter.

Hard cider's nutritional content relates to that of the apple juice from which it came, the type of fermentation, and any extra processing (such as filtration or pasteurization). Cloudy hard cider has a higher concentration of pectin and phenolic compounds than clear hard cider. For example, clear hard cider may have only 1–5 percent of the proanthocyanidin content of cloudy hard cider.

Hard cider contains relatively high levels of antioxidants: indeed, half a pint contains the same amount as a glass of red wine. Traditionally made natural unfiltered hard ciders have a higher level of those antioxidants called proanthocyanidins than do many red wines.

Lastly, hard cider contains fermentation products such as alcohol (derived from the apples' sugar) and small amounts of lactic acid (derived from the apples' malic acid, and which can add an interesting flavor). Hard cider is increasingly fashionable, but, however enjoyable it may be, it's wise to keep your alcohol intake within recommended limits.

Making hard cider

To make hard cider, cloudy or clear apple juice is either left to ferment naturally, or wine yeast is added to speed fermentation and make it more reliable. Shortly before fermentation ceases, the hard cider is siphoned off, leaving a sediment of dead yeast cells and other material at the bottom of the container. This is called "racking from the lees." Sparkling hard cider is made by allowing fermentation of the remaining sugar and, perhaps, adding more if necessary.

Most commercially produced hard cider is also treated in other ways. It is usually pasteurized – heated to 160ºF (71ºC) or treated with ultraviolet light to kill bacteria and molds. Unpasteurized hard cider can be risky for pregnant women, young children and people with poor immunity. But while pasteurization renders hard cider safer and increases shelf life, it can slightly alter its flavor. It also destroys enzymes and inhibits oxidation, giving a less distinct flavor.

What's more, much commercial hard cider is made from apple concentrate, contains artificial colorings, sweeteners, preservatives and enzymes, and is filtered. It may have a source of nitrogen added and be stored under compressed carbon dioxide gas. All this makes hard cider production more reliable and changes the color, clarity and flavor of hard cider in ways that some people prefer. Others prefer the appearance and flavor of simply made "natural" or "real" hard cider. To make your own hard cider, *see* pages 166–175.

Scrumpy can mean hard cider made from scrumps – or windfalls. Or it can mean cloudy and unsophisticated hard cider. Or it can mean "young" hard cider of a few months old that has not undergone maturation (which includes malolactic fermentation converting malic acid into lactic acid). Or it can mean fine hard cider made from choice apples, slowly fermented and matured for longer.

Apple Wine is usually made from dessert apples, so lacks sharpness ("bite," from fruit acid) and bitterness (from tannins). It often contains a higher concentration of alcohol too. The alcohol content of hard cider is nearly always less than 8 percent ABV, whereas apple wine is usually more alcoholic.

WHAT MAKES A REAL CIDER?

In the UK the Apple and Pear Produce Liaison Executive
(APPLE) aims to encourage appreciation of "real" cider.
They say top quality real cider must:

- be produced only from freshly pressed fruit
- not contain concentrate
- not be diluted
- not be pasteurized before or after fermentation
- not be fermented by added yeast
- not be treated with an enzyme
- not contain preservatives or coloring
- only contain sweeteners if labeled "medium" or "sweet," and then only if the sweetener is provably safe and does not affect the flavor
- not be filtered
- not have a nitrogen source added unless essential to start fermentation
- not be exposed to extraneous carbon dioxide

Apple Cider Vinegar

Making apple cider vinegar involves two types of fermentation. First, the sugars of apple juice are fermented to the alcohol of hard cider. Then the alcohol of hard cider is fermented to the acetic acid of apple cider vinegar (a process sometimes called "acetification").

The color of apple cider vinegar is a light brownish-yellow. It can be cloudy or clear. It can also be pasteurized (in which case it contains no microorganisms) or unpasteurized (in which case it contains some of the cloudy mass of fermentation bacteria called the "vinegar mother").

Although apple cider vinegar contains about 90 different substances, it is not a rich source of nutrients. One tablespoon, for example, contains a little carbohydrate, very small amounts of minerals, extremely tiny amounts of trace elements, and virtually no protein, fat, vitamins or fiber. Some people claim it's a good source of calcium, but it isn't. We need around 1,000mg of calcium a day from food. One tablespoon of apple cider vinegar contains only 1mg, whereas one tablespoon of milk contains 20mg. There is a paradox here, though, because many people, including around one in two over-60s, make insufficient stomach acid for optimal absorption of calcium and certain other minerals. Apple cider vinegar can increase stomach acidity for such people, in which case it increases the absorption of calcium from foods.

Other substances in apple cider vinegar include organic acids, flavonoids, pectin, enzymes and various volatile substances that enrich its scent and flavor. Apple cider vinegar has many health-giving properties. Many come from its organic acids (such as acetic and lactic acids and, perhaps, traces of malic acid). These acids are mainly responsible for its antifungal, antibacterial and antiviral actions. Malic acid, the main organic acid in apple juice, is fermented to milder

DID YOU KNOW?

Apple cider vinegar tablets are not at all well regulated and may contain no apple cider vinegar at all! Instead they may contain weak organic acid salts and flavorings that make them smell vinegary.

lactic acid during alcoholic and acetic fermentation. A by-product of this malolactic fermentation is a chemical with an attractive flavor that also contributes to the flavor of Chardonnay wine and some margarine spreads. Most commercial apple cider vinegars contain 5 percent acetic acid.

Apple cider vinegar adds to the stomach's natural acidity. It is absorbed from the gut into the bloodstream and almost completely oxidized in the body's cells to produce energy. Although it contains acids, its overall effect, once absorbed from the gut, is usually said to be slightly alkaline. Certain scientists explain this by saying that if vinegar is burnt in the laboratory to a dry ash, this is alkaline when tested with a pH (acid-alkaline) meter. They say the oxidation of vinegar's acids in cells to produce energy is equivalent to vinegar being burnt in the laboratory.

Good apple cider vinegar is aged slowly. Its flavor is enriched and its composition made more complex during fermentation and subsequent aging by volatile compounds such as aldehydes, ketones, alcohols, ethyl acetates, enzymes, phenolic compounds, salicylates and carboxylic acids (such as acetic, malic, lactic and succinic acids).

To make your own apple cider vinegar, *see* pages 175–177.

2

Natural
Remedies

The known health benefits of apples and apple products keep expanding as researchers discover more about their unique nutritional, health-giving and medicinal properties.

The saying, "An apple a day keeps the doctor away," is proving very true. Fruit makes an invaluable contribution to good health. Researchers recommend at least five helpings of fruit and vegetables a day; two of every five can be fruit; and one of these two can be fruit juice. However, many people fail to get anywhere near this five-a-day target. A survey by TNS Worldpanel UK in 2008 found only 12 percent of respondents had their five-a-day, while 12 percent had none at all.

Fruit is health-promoting and enjoyable to eat, and apples are especially so. They are very rich in certain substances – for example, flavonoids and pectin – that have important health benefits. A research review (*Nutrition Journal*, 2004) found apples were more consistently associated with a reduced risk of cancer, diabetes and heart disease than any other fruit (or vegetable). Apple consumption was also linked with less asthma, better lung function and increased weight loss. While apple juice and hard cider are less rich in health-giving ingredients than apples, they have some value and are delicious to drink.

As for apple cider vinegar, reports of its healing properties date from thousands of years ago. In 400 BC, the Greek doctor Hippocrates used it as an antibiotic and for general health. Samurai warriors used a vinegar tonic for strength,  and a vinegar solution was used to prevent stomach upsets and treat pneumonia and scurvy in the US Civil War and to treat wounds in World War I. But while traditional use, common sense and anecdotal evidence suggest that apple cider vinegar can help a wide variety of ailments, few trials have been done. One reason is that it's difficult to get funding for such research as apple cider vinegar cannot be patented.

Remember that you can also discourage common ailments with a healthy diet, adequate hydration, regular exercise, daily exposure to outdoor light, effective stress management, a sensible alcohol intake and no smoking.

Many of the most important diseases of our time, including diabetes and atherosclerosis (the furring and hardening of the arteries that encourages heart attacks and strokes), are associated with the sort of diet commonly eaten by obese people. Such a diet tends to have an unhealthy balance of fats and a lack of vegetables and fruit, both factors that encourage inflammation.

Research at the Universithy of Illinois, involving mice and reported in 2010, suggests that eating apples can prevent and treat these diseases. Specifically it found that soluble fiber, as in apples (and also, for example, in citrus fruit, strawberries, carrots, oats, nuts, seeds and lentils), increases production by immune cells of the anti-inflammatory protein interleukin-4.

People in the US are advised to eat 28–35g of total fiber a day. But this includes insoluble fiber. And although this sort helps move food through the gut, it doesn't boost the body's immunity.

Most food manufacturers do not include on a food's packaging the type of fiber it contains. It would be a great help to consumers to have new regulations requiring this information to be provided.

Cautionary Advice

Nothing should take the place of proper medical diagnosis and therapy. Check with your doctor before using apple cider vinegar with any existing medical treatment. And please bear in mind these side effects and safety issues:

- Anyone allergic to apple cider vinegar, apples or any apple constituent, such as pectin, should avoid them.
- The acidity of apple cider vinegar may temporarily soften tooth enamel, making it particularly vulnerable to damage and decay. So if you take apple cider vinegar as a food supplement or as a medicine, it's wise to dilute it in a glass of water, use a straw to drink it (so it bypasses your teeth) and rinse your mouth with plenty of water afterwards. If you need to clean your teeth, wait for at least 30 minutes.
- Excessive amounts of undiluted apple cider vinegar, or cider vinegar tablets, might damage the gullet and other parts of the digestive tract.
- A cider vinegar tablet may stick in the throat or gullet, so swallow it with a sip of water then wash it down well by drinking a further half a glass of water.
- A study in the US in 2005 found the ingredients of eight brands of cider vinegar tablet did not correspond with the details on packaging; analysis made researchers query whether they really contained only acetic acid.
- Prolonged use of apple cider vinegar could theoretically lower potassium, which could encourage toxicity from certain drugs (for example, digoxin, insulin, laxatives and certain diuretics).
- Apple cider vinegar affects blood sugar and insulin, so might have an additive effect if combined with diabetes medication.
- Apple cider vinegar may lower blood pressure, so it might have an additive effect if it is combined with high blood pressure medication.

Ailments A–Z

In this section I detail how and why apples, apple juice or apple cider vinegar might help in treatments, and whether this is based on traditional usage, anecdotal or medical evidence, or common sense.

Note that when I mention a study, I give the journal's name and year of publication; this plus some keywords should enable you to find out more via an internet search engine.

Acne

One cause of acne is overproduction of sebum due to over-sensitivity of sebaceous glands to testosterone. Other possibilities are changes in sebum and unusually sticky hair-follicle cells. Other triggers include the premenstrual fall in estrogen, humidity, stress, certain drugs (for example, the progestogen-only pill), and polycystic ovary syndrome. A reduction in the skin's normal acidity may encourage infected spots.

Some people report that apple cider vinegar helps; if so, it could be because it kills bacteria, increases skin acidity, "cuts" (emulsifies) skin oil, and reduces inflammation.

• Action:
Mix 1 part of apple cider vinegar with 4 parts of water. Apply with cotton wool, rinse after 10 minutes and repeat three times a day.

Age spots

The most common are brown freckles ("liver spots"), caused by normal aging plus photo-aging (accelerated aging from sun exposure).

Some people report that applying apple cider vinegar – particularly if mixed with onion juice – lightens age spots.

• Action:
Finely chop an onion, wrap it in muslin and squeeze to extract the juice. Mix 2 teaspoons of apple cider vinegar with 1 teaspoon of onion juice and apply the mixture to the freckles twice a day. They may begin to lighten within six weeks.

Aging

Scientists have long searched for lifestyle factors that encourage long life and discourage age-related diseases (such as arthritis, heart disease, diabetes, cancer, osteoporosis and Alzheimer's). Long-lived peoples include certain groups in Russia (the Georgians), Pakistan (the Hunzas), Ecuador, China, Tibet and Peru. One link is that they tend to live at high altitudes; here, melting glacier water is rich in alkaline minerals such as calcium, which help the body maintain a healthy pH (acid-alkaline balance) without drawing calcium from the bones. Apples, apple juice and apple cider vinegar are known

• Action:
Until we know more, hedge your bets by eating at least an apple a day, and either adding apple cider vinegar to various recipes, or taking 2 teaspoons in a glass of water 2 or 3 times a day.

Apples for longevity

A recent research project at the Chinese University of Hong Kong investigated the effects of apple polyphenols on fruit flies and found that they not only extended the average lifespan of the fruit flies by 10 percent but also helped them to retain good all-round general mobility.

Journal of Agricultural and Food Chemistry, 2011

to have an alkalinizing effect in the body, so they, too, can help to conserve stored calcium.

Another possible factor encouraging these peoples' longevity is their consumption of fermented vegetables, fruit, milk, cereal grains, meat or fish. We don't yet know whether any benefit is caused by fermentation reducing the carbohydrate in the food; by any remaining fermentation bacteria; or by the presence of fermentation acids (such as lactic or acetic). Whatever the reason, folk medicine has long held that apple cider vinegar (fermented apple juice) helps protect against age-related disease; certainly, consuming it before a meal discourages high blood sugar afterwards (*see* Diabetes). Also, people with an age-related reduction in stomach acid (this is believed to be one in two over-60s in Westernized cultures) who take apple cider vinegar, aid their absorption of many nutrients (including protein, carbohydrates, fats, vitamins A, B, C and E, calcium, iron, magnesium, zinc, copper, chromium, selenium, manganese, vanadium, molybdenum, cobalt).

US researchers say mineral and vitamin deficiencies can accelerate the age-related decay of mitochondria (energy-providing structures in cells). Among the most important deficiencies are those of iron, zinc, biotin, pantothenic acid, magnesium and manganese.

Molecular Aspects of Medicine, 2005

Apples can help reduce inflammation associated with heart disease, arthritis and Alzheimer's, because they contain antioxidants and aspirin-like salicylates with anti-inflammatory actions. Unpeeled apples contain larger amounts. Apple juice contains smaller amounts.

Apple consumption is also associated with a reduced risk of cancer, strokes and type 2 diabetes (*see* the research findings referred to in Chronic Illness).

Lastly, pectin binds to potentially toxic heavy metals such as aluminum and lead in the gut, which encourages their elimination. Indeed, pectin is regularly prescribed in Russia to remove heavy metals from the body. Such metals can form damaging "cross-links" with brain and other cells. So, pectin's binding ability may mean it helps protect against premature degeneration and aging.

DID YOU KNOW?

A single large apple with the skin has a higher value of antioxidants than an average half-cup serving of blueberries.

Alzheimer's Disease

This results from brain-cell destruction and is associated
with patches of amyloid protein and clusters of tangled
nerve fibers. The cause isn't clear, but it can run in families
and is more likely with age and after a serious head injury.

It's possible, though unproven, that eating apples
might help prevent Alzheimer's or slow its development.
One reason is that people with Alzheimer's tend to
have high levels of the amino acid homocysteine, and
are particularly likely to lack those B vitamins that help
normalize homocysteine levels. Apples contain small
amounts of folic acid and vitamin B6, which are among
the B vitamins with the greatest effect. So they can make
a useful contribution to the intake of vitamin B.

Several studies suggest that a small daily dose of
aspirin or other non-steroidal anti-inflammatory drug
discourages Alzheimer's. There isn't enough evidence for
doctors to recommend taking such drugs long-term, and
they can make the stomach bleed. But unpeeled apples,
and apple juice, hard cider and apple cider vinegar made
from unpeeled apples, are good sources of salicylates
with aspirin-like qualities.

What's more, research suggests that brain damage due
to the oxidation of the fatty acids arachidonic acid and
docosahexaenoic acid contributes to Alzheimer's – and
apples contain antioxidants.

• **Action:**
Eat an apple a day as this
will benefit your general
health and it might even
help prevent Alzheimer's.

Finally, studies suggest that increasing the amount of a neurotransmitter (nerve-message carrier) called acetylcholine in the brain can slow the mental decline associated with Alzheimer's. More recent research suggests that apples, and apple juice in particular, may help by increasing acetylcholine levels.

Laboratory tests in the US and Korea found that the antioxidant plant pigment quercetin helps protect rat brain cells from oxidation. Apple peel is rich in quercetin. So it's possible that eating unpeeled apples may help prevent Alzheimer's disease.

Journal of Agricultural and Food Chemistry, 2004

Researchers at the University of Massachusetts suspect that nutrients in apples and apple juice improve memory in mice, and protect against the oxidative damage that contributes to age-related brain disorders such as Alzheimer's. Consuming apple juice protected the mice from oxidative stress and slightly improved memory, possibly by increasing acetylcholine in the brain.

Journal on Nutrition Health and Aging, 2004

Anemia

Iron-deficiency anemia can be associated with low stomach acid, which affects one in two over-60s. It can also result from stress or the prolonged use of antacid or acid-suppressant medication. A lack of stomach acid can reduce iron absorption from food. Vitamin B12-deficiency anemia is another possible result of low stomach acid.

Eating apples could be particularly useful if you have iron-deficiency anemia. Gut bacteria break down apple pectin, releasing short-chain fatty acids which raise acidity and thereby boost iron absorption.

• **Action:**
If you have iron- or vitamin B12-deficiency anemia, try improving your absorption of either nutrient by drinking a glass of water containing 2 teaspoons of apple cider vinegar before each meal, or by adding apple cider vinegar to a dressing for a first-course salad or soup.

Anxiety

Eating apples might help reduce anxiety and panic attacks.

Researchers found that the urine of people with panic attacks was unusually acidic. Kidneys produce unusually acidic urine to help keep the blood's pH (acid-alkaline balance) within normal limits. One reason for overly acidic urine is an unbalanced diet with insufficient vegetables and fruit.

Psychiatry Research, 2005

Arthritis

Inflammation links the many sorts of arthritis. Although some people claim that apple cider vinegar helps their arthritis, others say it doesn't. While it's certainly possible that people might react in different ways, there is no scientific evidence, so the jury remains out.

Apples could help reduce the inflammation that often accompanies arthritis, because they contain antioxidants (such as proanthocyanidin plant pigments, beta carotene, vitamin C, selenium) and aspirin-like salicylates.

A folk remedy for arthritis in the hands or feet is to soak them 3 times a day in a solution made by adding 1 cup of apple cider vinegar to 3 cups of hot water.

Eat apples unpeeled to get larger amounts of natural anti-inflammatories. Apple juice contains smaller amounts and the best choice is cloudy juice made from crushed whole apples.

• Action:
Try eating more apples, other fruit, and vegetables for a month or so.

Asthma

Inflammation and oversensitivity of airways causes wheezing, coughing and a tight chest. Possible triggers include cold air, exercise, certain foods, hormone changes, laughter, infection, various fumes, a sudden fall in air pressure, thunderstorms, allergy, and breathing too fast.

Apples and apple juice have an anti-asthma effect that seems stronger than that of any other food. This may result from their high levels of antioxidants such as quercetin, as these have a provable anti-inflammatory effect.

• **Action:**
You may want to see if apple cider vinegar helps. Use it in salad dressings, soups or other recipes, or drink 2 teaspoons in a glass of water 3 times a day with meals.

Apples in pregnancy

Children of mothers who eat apples in pregnancy are much less asthma-prone in the first five years, say UK researchers. They attributed this to phenolic acids and other flavonoids.

Thorax, 2007

UK researchers reported that adults who ate at least two
apples a week decreased their asthma risk by up to a
third. They suggested that flavonoids in apples may
be responsible.

American Journal of Respiratory Critical Care Medicine, 2001

Another study found no protective effect from 3 subclasses
of flavonoids (catechins, flavonols and flavones), suggesting
that protection is due to other flavonoids.

European Respiratory Journal, 2005

A study of children with wheezing found that drinking
apple juice once a day was associated with improvement.

European Respiratory Journal, 2007

As for apple cider vinegar, Dr. D. C. Jarvis, who studied
24 people in the US over two years in the 1950s, found
their urine pH became highly alkaline before and early
on in an asthma attack (*Folk Medicine,* 1958). On following
his suggestion to drink apple cider vinegar, their urine
rapidly returned to its normal acid pH and the asthma
attack was less severe. He attributed this to its organic
acids and potassium content.

On searching for a clearer explanation, I came across a
1931 paper by Dr. George W. Bray of The Hospital for Sick
Children, London. He measured stomach acid after a meal

in more than 200 children with asthma. Astonishingly, 9 percent had no stomach acid, 48 percent had a severe lack, and 23 percent a slight lack. So 80 percent had an absence or deficiency of stomach acid. Other researchers have found a lack of stomach acid in many adults with asthma, as well as in people with other allergic conditions, including eczema, urticaria and hay fever. Dr. Bray also found that the blood of children with asthma was unusually alkaline. He suggested that their body's acid-producing ability was reducing the alkalinity of their blood instead of making stomach acid. He quotes an earlier study (*Journal of Physiology,* 1926–7) which suggests that sudden repeated exposure of sensitized tissue to a slight increase in alkalinity is extremely effective in triggering an exaggerated response. This implies that unusual alkalinity of blood and tissue fluid makes sensitized lung-lining cells more likely to become inflamed.

One possible cause of allergic sensitization is absorption from the gut of poorly digested protein, associated with low stomach acid. This work is extraordinarily interesting but has been largely ignored.

● **Action:**

Taking apple cider vinegar can restore some acidity to the stomach in someone with low levels of their own stomach acid. If you suspect this (for example, because you get indigestion that is not relieved by antacid medication), you might want to try taking apple cider vinegar either regularly each day to help prevent asthma, or early in an attack to help cure it.

For an adult, put 1 tablespoon of apple cider vinegar into a glass of water and sip over half an hour. Wait another half-hour then repeat. Or put 1 tablespoon of apple cider vinegar in soup or salad dressing. For children, use less apple cider vinegar, depending on their size.

Athlete's foot

This fungal infection makes the skin between the toes sore and soggy, and is often picked up in changing rooms or around swimming pools. Anecdotal reports suggest that apple cider vinegar might help.

• **Action:**
Each day bathe your feet for 5–10 minutes in warm water containing 4 tablespoons of apple cider vinegar and 10 drops of tea tree oil.

Bronchitis and emphysema

These two types of chronic obstructive lung disease often require ever more intensive treatment. Studies suggest apples may help. Most researchers believe that antioxidants are protective, and some think the flavonol quercetin plays a key role. Other constituents that may help include other flavonoids, pectins and malic acid.

• **Action:**
Eat an apple a day as this will benefit your general health and as studies suggest it might help prevent or ease bronchitis or emphysema.

A London study of 2,512 men found apples were the only food to benefit lung function. Volunteers eating five a week had a lung capacity 138ml (nearly ¼ pint, or up to 3 percent) higher than those eating none. This extra volume could make a big difference to "puffability." Apples seemed to slow the deterioration associated with smoking, and the authors thought that flavonoid antioxidants such as quercetin might be responsible.

Thorax, 2000

Researchers at the University of Nottingham found
evidence of fewer respiratory diseases in apple eaters.

Thoracic Society, annual meeting 2001

Bruises

These are associated with leaking of blood from tiny
blood vessels. Applying a solution of apple cider vinegar
to bruises is a folk remedy that could be worth trying to
limit bruising and speed recovery. If this works, it's very
possible that it does so because the flavenoids in apple
cider vinegar have a healing action. For example, they
are known to strengthen capillary walls and they also
enhance the capillary strengthening action of vitamin C.

DID YOU KNOW?

In Norse mythology, the goddess Iðunn
gave apples to the other gods to give them
eternal youth.

Cancer

Cancer results from mutation of a cell's DNA which allows the cell to continue multiplying instead of dying due to apoptosis (cell suicide) at its allotted time. Such malignant cells arise repeatedly during ordinary everyday life; most are destroyed by the immune system but a few grow into a cancer. Dietary factors that encourage certain cancers include a lack of antioxidants such as the flavonoid quercetin, as these normally mop up the free radicals (overactive oxygen particles) that are continually produced in the body and can damage cells.

Several studies suggest that apples have anti-cancer properties and that their antioxidants are partly responsible. Others point to other apple phytochemicals, including pectins, pectin-like rhamnogalacturonans, and triterpenoids. Certainly pectin is broken down by gut bacteria, releasing short-chain fatty acids which raise acidity in the large intestine and thereby encourage apoptosis in colon cancer cells. What's more, experiments indicate that pectins and pectin-like rhamnogalacturonans have pronounced antimutagenic effects.

Many studies have suggested that apples may help to prevent cancer, including:

Finnish research that followed 10,054 people from 1966 found cancer was least likely in those consuming most

• **Action:**
Eat an apple a day as this will benefit your general health and it might even help prevent cancer or slow its progression.

quercetin. The association between a reduced risk of lung cancer and apple consumption was especially strong.

American Journal of Clinical Nutrition, 2002

Evaluation of the antiproliferative activity of each of 13 apple-peel triterpenoids against liver, breast and colon cancer cells, indicates that each may contribute to the anti-cancer action of whole apples.

Journal of Agricultural and Food Chemistry, 2007

Researchers found that pectin inhibits galactoside-binding lectin 3 ("galectin 3"), a key protein in the progression and spread of cancer in the breast, prostate and colon.

Journal of the Federation of American Societies for Experimental Biology, 2008

Breast cancer

Researchers at Cornell University dosed rats with a breast-cancer inducer (mammary carcinogen) then gave half of them apple extract each day for six months. Tumor incidence was lower in those receiving the apple extract.

Journal of Agricultural and Food Chemistry, 2005

In a recent study at the same university, researchers found that apple extract slowed the growth ("proliferation") of adenocarcinoma breast cancers in rats; such tumors are the main cause of breast-cancer deaths. Consuming the equivalent in humans of one apple a day was associated with non-proliferation of their cancer in 43 percent of the rats; the equivalent of six apples a day doubled the likelihood of non-proliferation. In contrast, consuming no apple extract was associated with rapid tumor growth in 81 percent of the rats. Overall, in those rats given apple extract, the tumors were fewer, smaller, slower growing and less malignant.

Journal of Agricultural and Food Chemistry, 2009

Apples against cancer

When researchers at Cornell University in New York tested 25 fruits commonly eaten in the modern American diet, they found that apples are the biggest source of antioxidants, thanks to their high levels of phenolic compounds. They point out that increasing our antioxidant intake may reduce our risk of cancer.

Journal of Agricultural and Food Chemistry, 2008

Colon cancer

German tests found that adding apple pectin and phenolic compound-rich apple juice to stools encourages production of butyric acid – a short-chain fatty acid that inhibits the cancer-promoter histone.

Nutrition, 2008

Liver cancer

A study at Cornell University found that apples and especially apple peel have potent antioxidant activity and greatly inhibit liver-cancer cell growth.

Journal of Agricultural and Food Chemistry, 2003

Apple cider vinegar against cancer

Unlike other vinegars, apple cider vinegar has an alkalinizing effect in us. Our blood's normal alkalinity changes within only a very small range. However, certain experts believe being at the slightly less alkaline end causes "chronic metabolic acidosis" which raises our cancer risk. It's possible, though unproven, that apple cider vinegar as part of a healthy diet might help prevent cancer.

Lung cancer

A Hawaiian study of more than 10,000 people found a
40 percent lower risk of lung cancer in those who ate the
most apples.

Journal of the National Cancer Institute, 2000

Prostate cancer

Mayo Clinic researchers say that quercetin may help to
prevent or treat prostate cancer growth by blocking
androgen hormones.

Carcinogenesis, 2001

A University of Georgia study found that pectin triggered
apoptosis ("cell suicide") in up to 40 percent of prostate
cancer cells. It also affected cells that were resistant to
hormone therapy, and thus difficult to treat. Researchers are
trying to develop a drug that resembles the active part of
the most potent type of pectin.

Glycobiology, 2007

Some alternative practitioners believe that diet can
influence cancer by changing the body's overall pH (acid-
alkaline) balance. Certainly a cancer itself can be relatively
more acidic than normal tissue. But there is currently no
scientific evidence to support their view.

Cataract

This clouding of the eye's lens affects many people aged over 65. Apples might discourage cataracts by helping protect against certain trigger factors, including diabetes, high blood pressure, smoking, infection and sunlight. Their eye-friendly nutrients include vitamins B2 and C, flavonoids and salicylates.

Apple cider vinegar is an unproven folk remedy for helping to prevent cataracts worsening.

• **Action:**
If you would like to try apple cider vinegar, add it to recipes or take 2 teaspoons in a glass of water 3 times a day for 6 months.

Chronic illness

Eating an apple a day could indeed be a wise precaution because, for example:

A Finnish study that followed 10,054 people from 1966 onwards suggested that eating apples discourages many chronic diseases, including asthma, heart disease, cancer, strokes and diabetes.

American Journal of Clinical Nutrition, 2002

• **Action:**
Eat an apple a day as this will benefit your general health and might help prevent certain long-term illnesses.

Cold sores

These lip sores are caused by reactivation of a *Herpes simplex* viral infection by such triggers as stress, periods, skin damage, sunshine and fatigue. Apple cider vinegar is one of a number of traditional home-remedies for this condition. Others are listed in the box below.

• Action:

Try dabbing neat apple cider vinegar onto your lip 3 times a day with a paper tissue or clean cotton pad if you think a sore is imminent. You could also apply neat apple cider vinegar to an actual sore, but it might sting.

Alternative "cures" for cold sores

Apple cider vinegar's acidity may kill cold-sore viruses. Other folk remedies include applying petroleum jelly, lavender oil, tea-tree oil, coconut oil, oil from a vitamin E capsule, a used and cooled Earl Grey teabag, black coffee, or fresh lemon juice. It may also be well worth eating several cloves of garlic a day.

Colds and sore throat

Chewing an apple makes its pectin swell with water and form a soothing and protective layer of gel on an inflamed throat. This explains its use in certain commercial throat lozenges. An apple's vitamin C might shorten the length of a cold. Another benefit of apples is that fermentation of the pectin in the large intestine releases short-chain fatty acids (such as butyric acid) with "prebiotic" qualities – meaning they nourish "good" or "probiotic" bowel bacteria, such as lactobacilli and bifidobacteria. These, in turn, have beneficial effects on the body's immunity.

A German study of 479 volunteers confirmed previous studies in finding that taking probiotic tablets containing lactobacilli and bifidobacteria reduced the severity of colds, shortened their length by two days on average and increased immune cells such as helper T cells.

Clinical Nutrition, 2005

Dr. D. C. Jarvis of Vermont studied 24 people over two years in the 1950s and found their urine pH became highly alkaline before and early on in a cold (*Folk Medicine,* 1958*).* When they followed his suggestion to drink apple cider vinegar, their urine rapidly returned to its more usual acidic pH and the cold either didn't

• Action:

Put 1 tablespoon of apple cider vinegar in a glass of water and sip the mixture slowly over half an hour. Wait half an hour then repeat the treatment. If you dislike apple cider vinegar in water, add it to soup instead.

Gargle twice a day with a mixture made by putting 1 teaspoon of apple cider vinegar in a glass of water

develop or lasted only a short time. Dr. Jarvis attributed this to apple cider vinegar's organic acids and potassium. A sample of 24 is very small and I can find no other studies to back up this research. But you might in any case like to try taking apple cider vinegar in the early stages of a cold or sore throat.

A traditional remedy for a sore throat is to gargle with a dilute apple cider vinegar solution; why this might help is unclear, though apple cider vinegar does have some antibacterial properties.

Constipation

An unhealthy diet and dehydration are among the most likely causes of this condition. Eating unpeeled raw apples may help, because they are rich in water-soluble types of fiber called pectin and pectin-like compounds. Pectin dissolves in water to form a gel that makes stools softer and easier to pass through and from the bowel.

Unpeeled apples also contain cellulose; this insoluble fiber attracts water, which makes stools softer, bulkier and easier to pass, and reduces their "transit time" through the large intestine.

Cloudy apple juice contains smaller amounts of pectin than apples but may nonetheless be useful.

• **Action:**
Include apples and, perhaps, cloudy apple juice in your daily diet.

Apple juice and constipation

Apples contain sorbitol, an indigestible sugar alcohol that encourages diarrhea by attracting water, so can be helpful in treating constipation. A helping of apple juice supplies more than a helping of whole apple. Adults can usually handle apple juice without getting diarrhea. While apples may relieve constipation in a baby of six months or more, having more than 2 tablespoons twice a day may trigger an attack of diarrhea.

Corns and calluses

These are usually caused by ill-fitting footwear. Soaking the feet in a solution of apple cider vinegar on a regular basis is reputed to have the effect of softening corns and calluses and so hasten their demise.

• **Action:**
Add a cup of apple cider vinegar to a large bowl of warm water and soak your feet for 10 minutes a day. Afterwards, rub away the softened skin from the surface of the corn using a pumice or skin file.

Cough (*see also* Bronchitis and Emphysema)

Vinegar has been used for millennia to fight infections; Hippocrates (460–377 BC), for example, prescribed it for persistent coughs. Apples, too, may help to fight infections according to research.

Researchers found that a high consumption of fiber and fruit by people with chronic bronchitis was associated with a lesser likelihood of coughing. They believe flavonoids might be partly responsible – and whole apples are rich in flavonoids.

American Journal of Respiratory and Critical Care Medicine, 2004

A five-year study of the eating habits of 63,257 people in Singapore suggested that a diet high in fruit fiber reduces the likelihood of coughs.

American Journal of Respiratory and Critical Care Medicine, 2004

Traditional remedies for a cough involve applying some apple cider vinegar to the chest or pillow at night. By doing this small amounts of organic acid vapor from the vinegar might be absorbed into the body through the nose or skin, but why this might help remains unclear.

• Action:

Anyone prone to coughs could try eating 1 or 2 whole unpeeled apples each day.

Another idea – which might seem slightly whacky, but so what if it helps – is to soak some brown wrapping-paper in apple cider vinegar and put it on your chest. Cover it with a towel, and relax for 20 minutes. Alternatively, try sprinkling a little apple cider vinegar onto your pillow (covered with an old pillowslip!) each night.

Cramp

Possible triggers include insufficient dietary calcium, magnesium, potassium and vitamins B and C. These are all present in apples, so an apple a day might help.

Apple cider vinegar is a folk remedy for cramp, possibly because its acidity in the stomach improves calcium and magnesium absorption in the many people who produce sub-optimal levels of their own stomach acid.

• Action:

Include apple cider vinegar in recipes or take 2 teaspoons in a glass of water 3 times a day.

You might like to try this folk remedy too: mix 2 tablespoons of vinegar in a cup of warm water, soak a face flannel in this mixture, then put the flannel over the muscle and cover it with a thick towel.

Dandruff

This is often associated with the fungus *Malassezia furfur*. Apple cider vinegar is a popular home remedy as it is antibacterial and acts as an effective fungicide.

• Action:

Mix 1 cup of apple cider vinegar with 1 cup of warm water, apply this solution to the scalp, cover with a towel and wait an hour before rinsing and shampooing. Alternatively, massage neat apple cider vinegar into the scalp, cover with a towel and leave on for an hour before rinsing off. Repeat once or twice a week.

Diabetes and pre-diabetes

Apple cider vinegar and apples are proving useful in helping prevent or treat the high blood sugar of pre-diabetes and diabetes. This is important because high blood sugar encourages complications such as disease of the heart, eye and kidney.

A Finnish study of 10,000 people reported a lower risk of diabetes in apple eaters.

American Journal of Clinical Nutrition, 2002

Here are some examples of studies reporting that vinegar lowers blood sugar:

A study of 10 healthy volunteers at Lund University in Sweden found that including vinegar with a white-bread breakfast significantly reduced the expected rises in glucose and insulin afterwards. A vinegar breakfast also slowed absorption of paracetamol. The researchers attributed all this to the vinegar's acetic acid, and recommended eating fermented foods (such as vinegar) to reduce blood sugar and the need for insulin.

European Journal of Clinical Nutrition, 1998

• **Action:**

It seems sensible for people with pre-diabetes or diabetes to eat an apple a day and either add apple cider vinegar to their food (or drink 2 teaspoons of apple cider vinegar in a glass of water 3 times a day, with meals), or eat other fermented foods or pickled products containing vinegar.

Another study at Lund University found that eating pickled cucumbers with a white bread and yogurt breakfast dramatically lowered the expected spikes of blood sugar and insulin, while fresh cucumbers had no effect. They concluded this was due to the vinegar in pickled cucumber.

American Society for Clinical Nutrition, 2001

A study at Arizona State University involved 21 people with diabetes or pre-diabetes. Each drank water containing 2 tablespoons of apple cider vinegar before a carbohydrate breakfast. Vinegar increased insulin sensitivity by 34 percent in those with insulin resistance and 19 percent in those with

Blood sugar control

One reason why apple cider vinegar reduces the expected blood-sugar rise after a meal is because it suppresses the enzymes that break down sucrose, lactose and maltose. And just a small amount of apple cider vinegar diminishes the digestion of starch by an enzyme called salivary amylase (ptyalin).

diabetes. They were better able to get sugar from the blood into cells and their blood-sugar and insulin levels improved. Indeed, the blood sugar was 25 percent lower in those with diabetes, and nearly 50 percent lower in those with pre-diabetes.

Diabetes Care, 2004

A preliminary study of 12 healthy volunteers at Lund University found that taking vinegar with a white bread breakfast increased satiety; the bigger the dose, the greater the effect.

European Journal of Clinical Nutrition, 2005

A preliminary study at Arizona State University gave 11 people with diabetes 2 tablespoons of apple cider vinegar and a small piece of cheese before bedtime. Vinegar reduced pre-breakfast blood sugar by 6 percent next morning, showing its effect is long-lasting.

Journal of the Federation of American Societies for Experimental Biology, 2007

All this suggests that apple cider vinegar may aid blood-sugar control in people with diabetes, and slow the progression of pre-diabetes to diabetes. One possible explanation is delayed stomach emptying. Another, courtesy of Japanese researchers (*American Society for*

Nutritional Sciences, 2000), is that acetic acid inactivates intestinal enzymes (disaccharidases) that convert sugars to glucose. This would help prevent blood sugar rising too high or too quickly, so lowering the need for insulin. Other studies suggest that acetic acid helps normalize the release of sugar from the liver, and the production of sugar in the liver from non-carbohydrate sources.

Some studies show that other acidic foods, including lemon juice, yogurt, traditionally made slowly fermented bread, and kenkey (fermented corn) also reduce expected blood-sugar spikes after meals; some foods even being as powerful as the oral diabetes drug metformin. Frequent consumption of acidic foods is traditional in many

Apples and diabetes

An apple provides 20 percent of the recommended daily amount of dietary fiber. Because it's so rich in fiber, it helps control the blood-sugar level after a meal by slowing the rate of absorption of sugar into the blood. An apple a day as part of a healthy diet helps prevent pre-diabetes and keep diabetes well controlled.

countries and may help explain national differences in diabetes rates around the world. Many such foods contain acetic acid, including Japanese sunomono (vinegar-treated vegetables), sumeshi (vinegared rice), potato salad, mustard, vinegared fish and chips, and vinegar-based salad dressings.

As for apples, microorganisms in the colon degrade their pectin, liberating short-chain fatty acids such as butyric acid. These help prevent high blood sugar levels by reducing insulin release and inhibiting the breakdown of stored sugar (glycogen) in the liver. All this helps to prevent high blood sugar. Apple juice has less pectin, so raises blood sugar faster. Short-chain fatty acids may help in yet another way, because they inhibit C-reactive protein, a blood marker of inflammation and a predictor of diabetes.

Diarrhea

Apple pectin is water-soluble and in the gut it forms a gel that helps bind the bowel contents into stools and thereby reduces bowel-opening frequency. What's more, "good" bacteria in the gut break down some of the pectin, forming a protective coating for the gut lining which soothes any irritation. This breakdown releases short-chain fatty acids (such as butyric acid) with "prebiotic" qualities, meaning they nourish "good" or "probiotic" bowel bacteria such as lactobacilli and bifidobacteria. Pectin's prebiotic quality makes colon cells stronger and better able to produce protective mucus which helps prevent irritants sticking to and inflaming the gut lining.

Another type of apple fiber, cellulose, attracts water, bulks up bowel contents and makes them less runny. Cooking apples pre-softens their cellulose, helping it bulk up stools and reduce diarrhea.

Apple cider vinegar can help kill diarrhea-causing bacteria such as *Escherichia coli*, so it's useful for those people (such as one-in-two over-60s) with low production of stomach acid that would otherwise attack such bacteria.

• **Action:**

If you have diarrhea, try eating an apple every few hours. Raw apples are good, while cooking the apple first (for example, by stewing or baking it), pre-softens its cellulose, which may be useful if your bowel contents are rushing through very fast.

You might also want to try drinking 2 tablespoons of apple cider vinegar in a glass of cooled boiled water 3 times a day.

Diverticular disease

In this condition the colon is studded with "blowouts" called diverticula. This "diverticulosis" is usually symptom-free but can make the bowel irritable. The most common complication, diverticulitis, recurs whenever a diverticulum becomes inflamed. Diverticular disease becomes more likely with increasing age, inactivity and a poor diet that makes the colon unhealthy and encourages constipation.

Apples can be a great help because of their insoluble fiber (such as cellulose) and soluble fiber (such as pectin). Both types help prevent constipation. And the fermentation of pectin by "good" bacteria in the bowel releases short-chain fatty acids (such as butyric acid) which help nourish and protect colon lining cells.

• **Action:**
Eat an apple once or twice a day.

Ear infection

Apple cider vinegar is a traditional treatment for infection in the outer ear.

• **Action:**
Add 2 teaspoons of apple cider vinegar to an egg cup of water and apply with a cotton bud 3 times a day.

Eczema research

Research in Korea, reported in 2011, found that applying a vinegar-containing cream twice daily to the skin of mice with the sort of eczema called atopic dermatitis improved its appearance. The skin also became a more effective barrier and was more resistant to infection. This suggests that apple cider vinegar could help eczema in humans too.

Eczema

Apple cider vinegar washes are a traditional remedy for eczema. The inflammation of eczema makes the skin's pH (acidity/alkalinity) rise above its normal slightly acidic pH range of 4.2–5.6. Normal acidity helps prevent infection in eczematous skin by inhibiting the multiplication of potentially harmful bacteria and fungi. Apple cider vinegar washes have nearly the same pH as normal skin.

• **Action:**
Try rinsing the affected area of skin with a mixture of equal volumes of apple cider vinegar and water twice a day. Avoid applying to broken skin, as this will sting.

Fainting

Simple faints, or the dizziness that warns of them, are often associated with low blood sugar. Eating apples helps prevent this, mainly thanks to their content of the soluble fiber pectin, which helps keep the blood sugar steady by slowing the absorption of sugar from the gut.

Apple cider vinegar may help too, by slowing the rise in blood sugar after a meal; whether this is due to its acetic acid content or some other constituent is unclear.

Apple cider vinegar may also help prevent faints in those people – such as one in two over-60s – who have poor digestion caused by low stomach acidity, and who are also "fast oxidizers" of sugar, meaning they feel hungry sooner after a meal than do most people. This is because its extra acidity improves protein digestion, so they can readily produce energy from protein when they have used up their available sugar.

• **Action:**
Eat apples as between-meal snacks, to help maintain normal blood sugar. Include apple cider vinegar in your main meals in soups, dressings and sauces.

Fatigue

Apples supply sugar plus small amounts of B vitamins that may help prevent or treat fatigue. More importantly, they supply fiber, which helps keep blood sugar steady, so helping prevent the low-blood-sugar swings sometimes associated with fatigue.

• **Action:**
See if it helps to eat an apple a day, and to take 2 teaspoons of apple cider vinegar 3 times a day, as a condiment, in a glass of water, or added to recipes.

Theoretically, at least, apple cider vinegar could help tiredness associated with a lack of sufficient stomach acid. This prevents proper absorption of nutrients and is more likely with aging and stress.

Research at Nagoya University in Japan found that adding acetic acid – the main component of apple cider vinegar – to the diet of rats enhanced the replenishment of stored sugar (glycogen) in liver and muscle; this suggests apple cider vinegar might help prevent fatigue associated with endurance exercise.

Journal of Nutrition, 2001

US researchers claim that mineral and vitamin deficiencies accelerate age-related decay of mitochondria (the cells' "power-plants"). Among the most important deficiencies are those of iron, zinc, biotin, pantothenic acid, magnesium and manganese.

Molecular Aspects of Medicine, 2005

London researchers have developed a test (called the ATP profile) to indicate how well mitochondria are working. Their study of 71 people with chronic fatigue syndrome and 53 healthy controls suggested that mitochondrial dysfunction causes chronic fatigue syndrome/myalgic encephalomyelitis.

International Journal of Clinical Experimental Medicine, 2009

Food intolerance

Poor production of stomach acid can occur with
stress, tiredness, prolonged use of antacids or acid
suppressants and aging.

Normal levels of stomach acid enable the digestive
enzyme pepsin to break down proteins, but a shortage
allows poorly digested protein to be absorbed into the
blood and trigger allergy.

Alternative actions

*If you think you might be short of stomach acid, try
starting each main meal with a salad sprinkled
with an apple cider vinegar and olive oil
dressing, or first drink a glass of water containing
2 teaspoons of apple cider vinegar. Apple cider
vinegar is considerably less acidic than stomach acid,
but can nevertheless aid digestion.*

Gallstones

Most gallstones contain cholesterol, others contain bile pigments or calcium salts. The bile often contains excess cholesterol and the gallbladder doesn't contract well. Such problems are more likely with obesity, constipation or diabetes, all of which may be helped by consuming apples and/or apple cider vinegar as part of a healthy diet.

The apple fiber pectin may bind and thereby help eliminate certain bile acids from the gut, which would prevent them being reabsorbed and used in the production of gallstones.

A tendency to gallstones may be linked with a lack of stomach acid, because this encourages the gallbladder to be inactive, which in turn encourages the formation of gallstones in the stagnant bile. A lack of stomach acid is more likely with aging and stress and in people taking antacid or acid-suppressant medication.

Some people report success from a "gallbladder flush" (as described in the box opposite). This treatment is said to soften stones and let them come out in the stools next day. So far, very few success stories have been verified by x-ray or scan, and "softened gallstones" in stools may simply be lumps of soap formed from the salts and the oil.

IMPORTANT:

Before trying a flush, discuss it with your doctor.

• **Action:**

Eat an apple a day. Encourage gallbladder contractions with frequent meals that include something sour (such as apple cider vinegar) or bitter. If you suspect low levels of stomach acid, take 2 teaspoons of apple cider vinegar (added to a glass of water or a "starter") before a meal, to increase stomach acidity. Apple cider vinegar's pH is 5, whereas stomach acid is more strongly acidic, with a pH of 1–2, but a little extra acid might be useful.

A gallbladder flush

If your doctor agrees, you could try a gallbladder flush by drinking 1–2 liters of apple juice a day for 6 days. Next day, miss supper; at 9pm, take 1–2 tablespoons of Epsom salts in a little water; at 10pm, drink 4 fluid ounces of olive oil shaken with 2 fluid ounces of lemon juice, then lie on your left side for 30 minutes before bedtime. It may result in the stones being softened and excreted.

Gum disease

Chewing an apple boosts gum health, because repeated jaw movement increases the circulation of blood to the gums, and apples contain phenolic compounds called tannins which, some studies suggest, help prevent periodontal (gum) disease.

• **Action:**
An apple a day is good news for your gums!

Hay fever

Apple cider vinegar is a traditional remedy for allergic rhinitis. Dr. D. C. Jarvis, of Vermont, studied 24 people over two years in the 1950s and found that their urine pH became highly alkaline before and early on in an allergic attack (*Folk Medicine*, 1958). On following his suggestion to drink apple cider vinegar, their urine rapidly returned to a normal acid pH and the attack was less severe. He attributed this to the organic acids and potassium in apple cider vinegar.

I can find no confirmatory studies, but it should do no harm for adults to try apple cider vinegar early in an attack.

● **Action:**

Put a tablespoon of apple cider vinegar into a glass of water and sip the mixture over half an hour. Wait half an hour then repeat. Alternatively, add a tablespoon of apple cider vinegar to soup or other food.

Dr. Jarvis's "Honegar" cure

Dr. DeForest Clinton Jarvis, 1881–1966, was a physician whose book, Folk Medicine: A Vermont Doctor's Guide to Good Health *(1958), sold over one million copies. He believed that "honegar," a health tonic made from apple cider vinegar and honey, was sometimes more efficacious than medical drugs. He also advocated avoiding foods containing white flour and white sugar.*

Headache

Vinegar is a traditional remedy for a headache and familiar from the "Jack and Jill" nursery rhyme in which Jack mends his head with "vinegar and brown paper." Why vinegar should help isn't clear, but some complementary practitioners believe headaches can result from various body "buffer systems" having to work extra hard to keep the blood's pH (acid-alkaline balance) within its normal tightly controlled range; others believe headaches can result from this pH being at the alkaline end of normal. They recommend various remedies, a selection of which are listed, right.

• **Action:**
Sponge the head with apple cider vinegar, or apply a flannel soaked in a pint (600ml) of water containing 2 tablespoons of apple cider vinegar.

Inhale organic-acid vapor by putting a tablespoon of apple cider vinegar into a vaporizer and staying near for, say, 15 minutes. Or drink a cup of hot water containing 3 teaspoons of apple cider vinegar 3 times a day.

Head lice

Apple cider vinegar does not kill lice effectively but can loosen the glue that sticks louse eggs (nits) to hairs. Try the "Action" procedure described, right, and repeat it at frequent intervals.

• **Action:**
Add a cup of apple cider vinegar to a cup of water. Apply to dry hair then leave for half an hour. Wet the hair with water and smooth in lots of silicone-based conditioner. Comb the hair with a wide-toothed comb, then a fine-toothed one to remove any lice. Now shampoo. Some nits may stay stuck to hairs, so continue this wet-combing twice weekly for 2 weeks to catch newly hatched lice.

Heart disease

Coronary heart disease encourages angina and heart attacks. Fatty atheroma collects in coronary artery walls and, when overly thick, the heart muscle no longer gets enough blood to enable it to pump properly. Free radicals in blood oxidize LDL (low-density lipoprotein) cholesterol in atheroma, producing oxidized LDL cholesterol – the

• **Action:**
You could do your heart a favor by consuming apples and apple cider vinegar each day.

dangerous sort. Free radicals are overactive oxygen particles and are encouraged by a poor diet, infection, smoking and stress. They also trigger immune cells to inflame artery walls. Atheroma and inflammation scar and roughen the lining of arteries, which encourages high blood pressure by making arteries less elastic; they also encourage blood clots that can block an artery and cause a heart attack.

Studies suggest that apples and apple juice can help to protect against heart disease, possibly thanks to their flavonoids such as quercetin and catechins:

Finnish scientists who followed 5,133 initially healthy people for over 20 years found the risk of dying from heart disease was lowest in those who ate the most apples and other flavonoid-rich foods.

British Medical Journal, 1996

A study at the University of California Davis of 25 healthy people found that when they consumed 12 ounces of apple juice or 2 apples a day, their cholesterol took longer to oxidize. Slow-to-oxidize cholesterol is associated with a reduced risk of heart disease.

Journal of Medicinal Food, 2001

A US survey of nearly 40,000 women for almost 7 years reports that those who consumed apples had a risk of heart disease of up to 22 percent lower than those who did not eat apples. While this finding was not statistically significant, the researchers thought it warranted further investigation.

American Journal of Clinical Nutrition, 2003

A research review by doctors at Boston University School of Medicine suggests flavonoids improve the behavior of artery-lining cells and help prevent blood clots. This might help explain any beneficial effects on the cardiovascular disease risk.

American Journal of Clinical Nutrition, 2005

Analysis of 16 years of data on over 34,000 post-menopausal women in the Iowa Health Study, all free from cardiovascular disease at the start, found significant inverse associations between anthocyanidins and coronary heart disease mortality, cardiovascular disease mortality and total mortality; between flavanones and coronary heart disease mortality; and between flavones and total mortality. Apples were among those foods specifically associated with significant reduction in coronary heart disease and cardio-vascular disease mortality.

American Journal of Clinical Nutrition, 2007

Another possibility is that aspirin-like salicylates in apple peel may, like aspirin (acetyl salicylic acid), discourage heart attacks associated with inflammation of the coronary arteries.

Finally, "good" microorganisms degrade the apple fiber pectin in the large intestine, freeing useful short-chain fatty acids such as butyric acid. These acids reduce LDL cholesterol (the potentially dangerous sort) and increase HDL cholesterol (the potentially protective sort). They also inhibit C-reactive protein, a blood marker for inflammation and a predictor of cardiovascular disease.

As for apple cider vinegar:
Researchers at Harvard University who followed 76,283 women for 10 years found that women who consumed 1 to 2 tablespoons of oil-and-vinegar dressing most days had only half the average risk of heart disease. This salad dressing is a common source of alpha-linolenic acid, a polyunsaturated fat with known heart-health benefits, and the apparent benefit of oil-and-vinegar dressing was attributed to this. But it would be interesting to consider whether the vinegar also played a part.

American Journal of Clinical Nutrition, 1999

Heat rash

Apple cider vinegar is said to soothe this itchy pimply rash.

• **Action:**
Try applying a solution made by adding 1 tablespoon of apple cider vinegar to a cup of water.

Heavy periods

It's claimed that apple cider vinegar can ease heavy periods but there has been no scientific proof to date.

• **Action:**
If you would like to try this unproven home remedy, drink 2 teaspoons of apple cider vinegar in a glass of water 2 or 3 times a day, or add apple cider vinegar to your food.

Hiccups

Possible causes are an overfull stomach, due to eating too much; low stomach acid slowing protein digestion; or fatty, sugary food slowing stomach emptying and encouraging fermentation. Apple cider vinegar is a traditional remedy.

• **Action:**
To try preventing frequent attacks, add apple cider vinegar to your food, or drink a teaspoon of apple cider vinegar in a glass of water before a meal. To try stopping hiccups, very slowly sip the same solution or swallow a teaspoon of neat apple cider vinegar.

High blood pressure

Risk factors include obesity, overactivity of the kidney hormone renin, insulin resistance (pre-diabetes), salt sensitivity, age and genetics, though often there is no obvious cause. Vinegar can affect several of these factors and some studies suggest it can lower blood pressure.

There are five possible reasons. First, it increases nitric oxide (which relaxes blood vessels). Second, it acts like ACE-inhibitor blood-pressure medication (meaning it inhibits angiotensin-converting enzyme, thereby decreasing production of the blood-vessel constricting hormone angiotensin II). Third, it adds flavor, which helps salt-sensitive people reduce their salt intake. Fourth, adding it to casseroles or stocks containing meat bones releases some of their calcium; this could help the many people who have a low calcium intake, because calcium helps keep blood pressure healthy. Fifth, vinegar could lower blood pressure by encouraging weight loss.

Japanese researchers found that long-term dosage of rats with vinegar or acetic acid lowered blood pressure. They attributed vinegar's effect to its acetic acid. They believe this reduces renin activity, so decreasing angiotensin II. This, in turn, lowers blood pressure by decreasing blood volume and relaxing blood vessels.

Bioscience, Biotechnology, and Biochemistry, 2001

• **Action:**
If you would like to try apple cider vinegar, add it to your food 2 or 3 meals a day, or take a teaspoon in a glass of water 3 times a day.

Eat an apple a day.

Apples can help to lower high blood pressure by contributing potassium (which helps to regulate body fluids), magnesium (which relaxes blood vessel walls) and fiber.

High cholesterol

• **Action:**
Eat 1 or 2 apples a day.

Apples can help prevent high cholesterol. People with an unhealthy balance of LDL (low-density lipoprotein) and HDL (high-density lipoprotein) cholesterol tend to develop a cholesterol-rich layer of atheroma in their arteries. This impairs their circulation. Also, the presence of oxidized LDL cholesterol in atheroma stiffens arteries and encourages high blood pressure, heart attacks and strokes.

Pectin in apples, and to a lesser extent in cloudy apple juice, absorbs cholesterol and triglycerides in the gut and eliminates them from the body. One way it does this is by increasing the viscosity of the contents of the small intestine, which reduces the absorption of cholesterol from food or bile. Another is that "good" microorganisms degrade pectin in the large intestine, liberating short-chain fatty acids (such as butyric acid) which inhibit cholesterol absorption, suppress cholesterol production in the liver, and boost HDL cholesterol. Studies using whole apples show that a combination of pectin and vitamin C lowers cholesterol more than does pectin alone; and the combination of pectin and phenols lowers cholesterol and triglycerides more than either alone. The decreases are small but worthwhile. Eating one large apple a day lowers cholesterol by up to 11 percent. Eating two lowers cholesterol by up to 16 percent. The cholesterol-lowering effect of four a day can equal that of a statin drug!

In a study at the University of California Davis, volunteers who consumed 2 apples or 12oz of apple juice a day demonstrated significant slowing of cholesterol oxidation. The protective effect reached its peak after 3 hours and dropped off after 24, backing the oft-repeated advice to "eat an apple a day."

Journal of Medicinal Food, 2000

Japanese researchers found that giving rats acetic acid (as in vinegar) lowered their cholesterol and triglycerides. Further tests revealed inhibition of production of triglyceride fats (now properly known as triacylglycerols) from sugar in the liver, and an increase of bile in the gut. We don't yet know if vinegar lowers blood pressure in humans, or, if it does, how.

British Journal of Nutrition, 2006

Some alternative practitioners explain that cholesterol is an acidic by-product of fat metabolism. They say that a healthy diet containing plenty of alkaline-forming foods (such as apples and apple cider vinegar) makes the body better able to prevent atheroma accumulating in the arteries, and to go on to dissolve or neutralize and then eliminate cholesterol.

Indigestion and heartburn

Stewed apple is a favored first food after a gastrointestinal infection. And organic acids in apples (for example, malic and tartaric) and apple cider vinegar (acetic acid) may help prevent indigestion caused by low stomach-acid production. Low stomach acid is encouraged by aging, stress and prolonged use of antacids or acid-suppressant medication. Organic acids such as those in vinegar are weaker than gastric acid, but help provide an acidic environment for efficient protein digestion. Finally, pectin in apples and cloudy apple juice helps keep the gut free from sticky residues which make it sluggish. It also helps prevent potentially harmful bacteria binding to the gut lining.

Heartburn is a frequent symptom of low stomach acid. Medical treatment is to suppress gastric acid with antacids, but this sometimes does no good or even worsens the problem. If antacids don't help, you may have low stomach acid. Consuming apple cider vinegar could then help by

• **Action:**
Eat an apple a day.

To use apple cider vinegar to prevent problems, take 2 teaspoons in a glass of water before each meal, or add it to food.

If you have indigestion or heartburn, take 1 tablespoon of apple cider vinegar in a glass of water. If it helps, the odds are that a lack of stomach acid was to blame.

increasing your stomach acidity, though because its pH
is only around 5, it cannot make the stomach as acidic as
gastric acid (pH 1–2).

Irritable bowel syndrome (IBS)

Possible symptoms include pain, constipation, diarrhea,
passing mucus, bowels never feeling empty, wind and
bloating. One in three people sometimes have an irritable
bowel; one in five of these have frequent trouble – which
is then called irritable bowel syndrome (IBS).

Apples' soluble pectin fiber may reduce symptoms,
partly because it makes stools softer and easier to pass.

• Action:
Try eating an apple a day.

Itching

Applying apple cider vinegar topically is said to help
relieve itching.

• Action:
Try bathing in tepid bath
water containing a cup
of apple cider vinegar. Or
try applying neat apple
cider vinegar to itchy
skin, keeping it away
from your eyes or other
delicate parts.

Kidney stones

Apple cider vinegar is said to help dissolve common, calcium-containing stones. These are more likely when threatened overacidity of body fluids leads to calcium being withdrawn from bones and teeth and excreted in the urine to keep the body-fluid's pH (acid-alkaline balance) within its normal tightly controlled range. Apple cider vinegar, unlike other vinegars, is said to have a mild alkalinizing effect after it has been digested, so it might reduce the body's need to take calcium from bones. It might also help by reducing spikes of insulin in the blood after eating carbohydrate. Such foods otherwise raise insulin – a hormone that encourages stones by making the kidneys discharge more calcium in the urine.

Apples and apple juice may help prevent or dissolve stones, as they too have an alkalinizing effect. They also provide vitamin B6 and magnesium; this might help because, according to researchers, a lack of these nutrients encourages stones to develop.

• **Action:**
Try including apple cider vinegar, apples and apple juice in your diet if you are prone to kidney stones.

Low immunity

Apples may boost immunity. First, their vitamin C
content is useful for immunity. Second, their pectin
fiber is degraded by "good" microorganisms in the large
intestine, liberating short-chain fatty acids such as butyric
acid. These aid immunity by stimulating production in
the spleen of helper T cells, antibodies, white blood cells
and cytokines. They also inhibit C-reactive protein, a
blood marker of inflammation.

• **Action:**
Eat an apple a day.

DID YOU KNOW?

The apple is the official state fruit of New York,
Rhode Island, Vermont, Washington, and
West Virginia.

Memory loss

Anecdotal reports suggest that apple cider vinegar can aid memory.

As for apples and apple juice:

Researchers at the University of Massachusetts suspect that nutrients in apples and apple juice improve memory and learning in mice, and protect against the oxidative damage that contributes to age-related brain disorders such as Alzheimer's. Consuming apple juice protected against oxidative stress and slightly improved memory and learning, possibly by increasing the brain neuro-transmitter (nerve-message carrier) acetylcholine. The amount was comparable to humans drinking two 8oz glasses of apple juice or eating 2 to 3 apples a day.

Journal on Nutrition, Health and Aging, 2004

• Action:

There is only very preliminary evidence that apples, apple juice and apple cider vinegar can improve memory or slow its loss, but there's absolutely nothing to be lost by including them in your diet.

Metabolic syndrome

This collection of symptoms greatly encourages diabetes, heart disease and strokes. It affects one in five of us overall and is also known as insulin resistance syndrome and syndrome X. It's more likely with increasing age and can be associated with polycystic ovary syndrome. Diagnosis is based on having some combination (subject to debate) of high fasting blood sugar (pre-diabetes), high blood pressure, an apple-shaped body, low HDL-cholesterol and high triglyceride blood fats. Most people who have this syndrome are sedentary, obese and insulin-resistant, though it's unclear whether obesity and insulin resistance are causes or consequences of a more general problem. Some researchers think inflammation and oxidation are involved. Certainly, affected people are more likely to have a raised level in the blood of C-reactive protein, which indicates inflammation.

The US National Health and Nutrition Examination Survey (1999–2004) revealed a 27 percent decrease in risk of metabolic syndrome among regular consumers of apples, apple sauce and apple juice. They also had a 30 percent lower likelihood of high diastolic blood pressure; a 36 percent lower likelihood of high systolic blood pressure; a 21 percent lower likelihood of a large waist; and lower levels of C-reactive protein, an indicator of inflammation.

Experimental Biology meeting, 2008

• **Action:**
It makes sense to eat an apple a day just in case it is protective.

Nosebleed

There are isolated reports that apple cider vinegar helps
to stem a nosebleed. Also, consuming apple cider vinegar
is a traditional remedy for frequent nosebleeds.

Alternative actions

*To see if apple cider vinegar helps stop a nosebleed, soak a cotton-
wool ball in apple cider vinegar, lean your head backwards, then put
the cotton-wool ball in the affected nostril. If you suffer from frequent
nosebleeds, try drinking 2 teaspoons of apple cider vinegar in a glass of
water 3 times a day over a period of time, say, 3 months.*

Osteoporosis

In this condition, affected bone is light and fragile and its cells are destroyed faster than they are created. Risk factors include age, too much or too little exercise, smoking, too little bright outdoor light, a lack of bone-friendly nutrients (calcium, magnesium, zinc, vitamins C, D and K and plant hormones), an early menopause, anorexia and various medications and illnesses (including gut and thyroid disorders). Research increasingly points to inflammation and oxidation being involved.

Apples might help prevent osteoporosis or slow its development because of their antioxidants such as flavonoids, which counter inflammation and oxidation. Apples also contain the trace mineral boron, which researchers believe could improve estrogen levels and reduce the loss of bone-friendly minerals in the urine. Another constituent, present in small amounts, is the phyto-estrogen genistein. The rate of loss of bone density around the menopause is much lower in women with a high intake of plant estrogens; apples contain only small amounts, nevertheless they might help as part of a healthy diet. Apple pectin is useful too, because good gut bacteria break this down, releasing short-chain fatty acids which raise acidity in the large intestine and thereby boost absorption of minerals such as calcium and magnesium.

• **Action:**
It's worth including apples and apple cider vinegar in your daily diet to help prevent or treat osteoporosis.

Apple cider vinegar might be useful too. First, it renders calcium from food or in supplements more soluble and thus better absorbed. Second, it provides extra acidity in the stomach, which is useful for people in whom a lack of stomach acid causes poor absorption of calcium and certain other nutrients. Third, although apple cider vinegar is acidic, the net result of its digestion and metabolism is mildly alkaline.

French researchers report that the phenolic compound phlorizin, found only in apples, prevents bone loss associated with inflammation in rats. If true in humans too, then eating apples might help prevent or treat osteoporosis.

Calcified Tissue International, 2005

DID YOU KNOW?

The skin of an apple contains more antioxidants and fiber than the flesh.

Overweight and obesity

Apples contain the soluble fiber pectin, which slows the absorption of sugar from the gut. This helps prevent hunger and overeating. Pectin can also interfere with the absorption of fat. One theory is that this is because pectins form a gel in the stomach, which mops up triglyceride fats and stops them being absorbed.

Researchers in Texas found that taking pectin increased satiety.

Journal of the American College of Nutrition, 1997

Researchers in Rio de Janeiro reported that overweight middle-aged women on a weight-loss diet who ate 3 apples or pears a day lost more weight.

Nutrition, 2003

Apple cider vinegar has been said for thousands of years to promote weight loss. Early research suggests that if apple cider vinegar does indeed help, it does so by aiding satiety after eating, by helping the body burn calories faster and by helping to compensate for any lack of stomach acid.

Low acidity in the stomach affects around one in two over-60s, and is encouraged by stress and prolonged use of antacid or acid-suppressant medication. Research associates low stomach acidity with poor absorption of many nutrients (including protein, vitamins B and

• **Action:**
It's worth including apples and apple cider vinegar in your diet if you would like to maintain or achieve a healthy weight.

C, calcium, iron, magnesium, zinc, copper, chromium, selenium, manganese, vanadium, molybdenum and cobalt). It also indicates that poor nutrient absorption can make people want to eat even when they are not hungry. Apple cider vinegar improves absorption of these nutrients by increasing stomach acidity. And by improving protein digestion, it may specifically help people who are "fast oxidizers" of sugar and tend to feel very hungry within three hours or so of a meal. This is because they can readily produce energy from protein when they have used up their available sugar.

Apple cider vinegar also aids absorption of fats and vitamins A and E by stimulating the release of bile and pancreatic enzymes into the gut. It also slows the rise in blood sugar after a meal. This not only helps prevent high blood sugar, but also the low-blood-sugar swing that sometimes follows, and which can trigger the desire to overeat. Whether its blood-sugar-lowering ability is due to its acetic acid or another constituent is unclear.

Japanese research suggests that acetic acid lowers blood sugar by reducing the activity of disaccharidases (enzymes which break complex sugars into simple sugars prior to absorption).

Journal of Nutrition, 2000

A study of 18 people at Southern Utah University examined how acids in vinegar, pickle juice, acetic acid, lemon and lime juice, and a placebo, affected the expected blood sugar after a carbohydrate snack. Only vinegar significantly lowered it, which suggests that compounds in vinegar other than acetic acid must contribute to the effect.

Journal of the Federation of American Societies for
Experimental Biology, 2006

Apples as an aid to weight-loss

Women aged 45–65 who eat 75g a day of dried apple had an average weight loss of 3.3lb in a year, says a study at Florida State University, presented in 2011. Also, their cholesterol balance improved and their inflammation markers decreased. This is probably because apple pectin encourages satiety, and apple polyphenols improve fat metabolism and discourage inflammation.

Experimental Biology meeting, 2011

When 12 volunteers took 2 tablespoons of vinegar before a carbohydrate-rich meal, they had lower than expected levels of glucose and insulin, and their feeling of fullness more than doubled.

European Journal of Clinical Nutrition, 2005

A report claims that women in their 50s gained 5 pounds less over 10 years if they took more than 500mg of calcium supplements than if they did not. This suggests that people with low stomach acid might lose weight more easily if they improve their calcium absorption by taking vinegar with meals.

On-line report of work at the
Fred Hutchinson Cancer Research Center in Seattle, 2006

Japanese researchers report that mice given both a high-fat diet and acetic acid each day for six weeks gained a lot less body fat than did mice given a high-fat diet but no vinegar. They also noted that the acetic-acid group had increased activation ("upregulation" or "turning on") of genes which produce proteins that break down fats. These findings lead them to believe that acetic acid activates genes that break down fats.

Journal of Agricultural and Food Chemistry, 2009

Parkinson's disease

This results from degeneration of brain cells that produce the neurotransmitter (nerve-message carrier) dopamine. There's usually no apparent cause but scientists believe genes and brain infection can play a part.

Studies suggest that an apple a day helps reduce the risk of neurodegenerative disorders such as Parkinson's.

A study at Cornell University in the US compared how 2 groups of rat nerve cells fared against hydrogen peroxide, a common oxidator. One group was pre-treated with apple phenolic extracts; the higher the concentration of the apple extract, the greater the protection against oxidation.

Journal of Food Science, 2004

A study at the same university found the antioxidant quercetin seemed mainly responsible for the above protection, and was better than vitamin C at protecting nerve cells. Apples are rich in quercetin.

Journal of Agricultural and Food Chemistry, 2004

• **Action:**
Try eating an apple a day.

Peptic ulcer

Thick mucus normally protects the stomach and duodenum from stomach acid and the digestive enzyme pepsin. An ulcer can develop if something interferes with this mucus or with the lining cells or the volume of acid. The usual culprit is inflammation from infection with *Helicobacter pylori* bacteria. This is a major cause of stomach ulcers, stomach inflammation (gastritis) and stomach cancer. Around 2 in 5 of us are infected, though only 1 in 10 infected people develop an ulcer. Some people with ulcers make too much acid, but most don't, and some make too little.

Japanese research shows a strong correlation between low stomach acidity and increased rates of *H. pylori* infection.

Biotechnic and Histochemistry, 2001

It's theoretically possible that if someone with peptic ulcer symptoms tests positive for *H. pylori* and suspects a lack of stomach acid (for example, because acid-suppressants don't relieve symptoms), the acidity of apple cider vinegar might discourage the infection. But it might temporarily worsen ulcer pain.

• **Action:**

If you would like to try it, either drink a glass of water containing 2 teaspoons of apple cider vinegar each day, or use apple cider vinegar as a condiment or add it to suitable recipes.

Piles

These painful swollen veins in the lining of the back passage are often associated with constipation.

• **Action:**
Including apples in your diet could make all the difference to the problem.

Smelly feet

Apple cider vinegar soaks are said to reduce foot odor for some hours.

• **Action:**
Soak your feet when necessary in a bowl of hot water plus a cup of apple cider vinegar.

Sprains

Apple cider vinegar is said to relieve the pain from a sprain, though why is unclear.

• **Action:**
Apply an apple cider vinegar compress to the affected area: for example, a face cloth squeezed out in a bowl of hot water with a cup of apple cider vinegar added.

Stings

Applying apple cider vinegar is a traditional way of treating a wasp sting. A bee sting requires bicarbonate of soda – baking soda – instead. An easy way to remember which treatment is required is "V" for "vasp" stings and vinegar, "B" for baking soda and bee stings. One possible mechanism is that vinegar can convert certain toxins in wasp venom to less toxic acetate compounds.

Dousing with vinegar is a folk remedy for most jellyfish stings as it deactivates venom cells. But immersing the stung part in hot water, if available, for four minutes, is even more effective. Don't put vinegar on a sting from a Portuguese man-of-war jellyfish, though, since researchers say this could make the venom cells that are embedded in the skin discharge more venom.

• **Action:**
Apply neat apple cider vinegar to a wasp or jellyfish sting, using a cotton pad.

Strokes

A stroke ("brain attack") usually results from a blood clot interrupting the blood flow in one of the brain's blood vessels (thrombotic stroke). Less often it is caused by bleeding in the brain from an unhealthy artery (hemorrhagic stroke).

The main culprit behind a thrombotic stroke is the narrowing of an artery by atheroma. This fatty substance contains low-density lipoprotein cholesterol, which is readily oxidized by free radicals, making arteries inflamed, scarred and rough inside. Clots form on roughened artery walls, especially if blood is abnormally sticky. Risk factors include smoking, stress, unhealthy diet, obesity, high blood pressure, diabetes and chronic infections.

A Finnish analysis of the diet and health of 9,208 men over a period of 28 years found the lowest risk of thrombotic stroke in those who ate most apples.

European Journal of Clinical Nutrition, 2000

Preliminary evidence suggests that apples and apple cider vinegar help prevent high blood pressure, obesity and diabetes, so it's worth adding them to your diet.

• **Action:**
Eat at least 1 apple a day and include apple cider vinegar in your daily diet.

Ulcerative colitis

Consuming apples might help ulcerative colitis, thanks to their high content of the soluble fiber pectin.

People with mildly to moderately active ulcerative colitis became less reliant on other therapies when they took a supplement of soluble fiber, fish oil and antioxidants.

Clinical Gastroenterology and Hepatology, 2005

• **Action:**
Eat an apple a day.

Varicose veins

Applications of apple cider vinegar are a traditional remedy for soothing aching varicose veins. If the aching is associated with phlebitis – meaning inflammation of the lining of the swollen veins – it's possible that traces of polyphenols and other antioxidants from apple cider vinegar penetrate the skin and the vein walls and then exert a soothing influence.

Consuming a little apple cider vinegar each day (as described right) may help prevent varicose veins by encouraging weight loss.

• Action:
Dampen a small towel with apple cider vinegar and wrap it over troublesome veins twice a day, for half an hour each time.

Also, either drink a glass of water containing 2 teaspoons of apple cider vinegar 3 times a day, or use apple cider vinegar as a condiment or in recipes.

Warts

There are many anecdotal reports of apple cider vinegar curing warts, but I can find no scientific evidence supporting its use. Some podiatrists use a more corrosive relative of acetic acid, dichloroacetic acid, to treat verrucas and other warts.

• Action:
Soak some cotton wool in apple cider vinegar and apply to the wart. Cover with a sticking plaster overnight. Repeat each night for 2 weeks.

3

BeautyAid

As well as improving your health and well-being, apples and apple cider vinegar can bring a glow to your face and hair.

Apples and apple cider vinegar are not only valuable for our health but also have a rightful place in the beauty salon. The main reason that apple cider vinegar is such a popular beauty aid is that its organic acid concentration of about 5 percent helps maintain the skin's natural acidity. Most other vinegars – except, for example, naturally fermented wine vinegar – are more acidic than this and therefore unsuitable for skin care.

Normal skin has a slightly acidic surface layer called the "acid mantle" or hydrolipid film. This contains:

- The fatty acids of skin oil (sebum)
- Lactic acid and various amino acids from sweat
- Amino acids and pyrrolidine carboxylic acid from "cornifying" (hardening) skin cells.

DID YOU KNOW?

Apples and apple cider vinegar are rich in alpha hydroxy acids (AHA) which feature in many expensive skin-care products.

The skin's acid mantle has a pH (acid/alkaline balance) of 4.5–5.75 over most of the body. (A "neutral" pH is 7; below this is acidic, above is alkaline). The pH of the skin in the armpits and around the genitals is around 6.5, which is less acidic. Normal skin pH tends to be slightly more acidic in men than in women.

Normal acidity helps activate the enzymes that enable the production of the lipids (oily fats) present in the skin's hydrolipid film. It also encourages skin to repair itself after mechanical or chemical damage. All this is important, because intact healthy skin is relatively impermeable. This means that water is much less able to escape through the skin (other than via perspiration), and potentially harmful substances and microorganisms are less able to get in. Normal skin acidity also encourages the development of a normal skin flora – the typical populations of various bacteria and fungi that inhabit healthy skin. A normal skin flora helps to prevent potentially harmful microorganisms from multiplying and causing infections.

Any loss of normal skin acidity encourages drying, cracking and itching. What's more, eczema or other inflammation tends to make skin more alkaline. Washing with most types of soap increases this alkalinity and makes the skin even more vulnerable to irritation and infection.

Most soaps, even "mild" soaps, glycerine soaps, "baby soaps" and "beauty bars," have an alkaline pH of 7–9. Washing with such soap destroys the skin's protective acid mantle. Healthy, unbroken skin can recover from this increase in pH but the restoration of normal acidity takes time – generally between half an hour and two hours or more; and twice-daily washing with alkaline soap slightly reduces the restored acidity level. Certain soaps are even more alkaline, with a pH of 9.5–11, so they compromise skin acidity even more. Look out for soaps with a comparatively low pH balance (7.5 or less). Only a very few bar soaps have a pH similar to that of normal skin.

However, the pH of many liquid soaps, non-soap cleansers and bath and shower gels is closer to that of normal skin; and a few have a pH similar to that of normal skin (around pH5). Using a homemade skin cleanser containing apple cider vinegar avoids the loss of normal acidity that accompanies washing with most types of soap. Another idea, if you want to continue using alkaline soap,

DID YOU KNOW?

The antibacterial properties of apple cider vinegar make it particularly effective for deodorizing and refreshing armpits and feet.

is to rinse your skin afterwards with a homemade skin splash containing apple cider vinegar, so as to restore the skin's normal acidity.

Apple cider vinegar can also restore acidity to hair that has been washed with an alkaline shampoo. Most shampoos are alkaline and can leave newly washed hair looking dull and lackluster. They also temporarily destroy the normal acidity of the scalp, leaving it more prone to dryness, irritation and infection. However, an apple cider vinegar rinse can soothe the scalp, get rid of any residue build-up of hair products and make the hair shinier than it would otherwise be. It can also enhance the hair's natural highlights.

Skin cleansing

- Make a cleansing rub for soiled hands by moistening half a cup of oatmeal with apple cider vinegar. Use a larger quantity of oatmeal if you want to cleanse your whole body this way.

- Put a cup of apple cider vinegar in the bath water, immerse a wash-cloth in the water and use it to cleanse your skin.

Skin rinsing

- Wash in a shower or unplugged bath and rinse yourself with water. Then fill the bath with water, add half a cup of apple cider vinegar, lie back and relax.

- Alternatively, wash in the bath or shower. Then rinse yourself with warm water and half a cup of apple cider vinegar poured from a plastic pitcher.

Skin toning

- If you usually use cleansing cream or lotion on your face, follow this by applying a skin toner made by adding 4 tablespoons of apple cider vinegar to half a pint (250ml) of cold water. Keep the skin toner in a capped glass or plastic bottle, and apply it with a soft cotton cloth or cotton wool.

- Alternatively, use a scented toner. Make the toner as above, put it in a saucepan and add half a tablespoon of dried rosemary leaves or lavender flowers, or 1 tablespoon of fresh leaves or flowers. Bring to a boil and simmer for 5 minutes, then cool, strain and bottle.

Hair rinsing

- After shampooing, rinse your hair with a 1-pint (500-ml) pitcher of warm water to which you have added 2 tablespoons of apple cider vinegar.

Deodorizing

- Wash your armpits, and then apply either neat apple cider vinegar, or apple cider vinegar in which you have steeped some rosemary or mint leaves, or lavender flowers, for 2 weeks.

- Wash your feet, then soak them for 5–10 minutes in a basin of water containing half a cup of apple cider vinegar.

4

Household Help

Apple cider vinegar can make household chores easier and greener while saving the expense of buying household maintenance products.

Air-freshening

- Half fill a spray bottle with water and add a teaspoon of baking soda (sodium bicarbonate) and a tablespoon of apple cider vinegar. Shake the open bottle to mix the contents. When the foaming stops, fill the bottle with water and put on the cap. Use as an air freshener to help eliminate the smell of smoke, pets, cooking and other unwanted odors.

- Make a fragrant freshener by adding to a saucepan 2 pints of water, 2 tablespoons of apple cider vinegar and 2oz (50g) fresh, or 1oz (25g) dried, lavender, thyme, rosemary or cloves. Bring to a boil and simmer for 10 minutes. Cool, strain, put into a spray bottle and use as desired.

- Rid your hands of the lingering smell of onions, garlic or fish by pouring a little apple cider vinegar into your cupped hand, rubbing your hands together, and then washing with soapy water.

- Add a splash of apple cider vinegar to soapy water then use this to wipe kitchen worktops or other hard surfaces to rid them of food smells or other unwanted odors.

- When painting a room, reduce the paint odor by standing a bowl of apple cider vinegar somewhere safe.

Cleaning and laundry

- Rub soiled collars and cuffs with a paste of equal parts cider vinegar and baking soda (sodium bicarbonate). Leave 30 minutes then wash as normal.

- Reduce perspiration stains on clothing by soaking garments for several hours in a basin of water containing half a cup of apple cider vinegar.

- Freshen and clean floors by adding a cup of apple cider vinegar to the cleaning water.

- Help to keep sink and basin drains clear by pouring half a cup of baking soda (sodium bicarbonate) down the plughole, then half a cup of hot apple cider vinegar (heated for a minute in the microwave). Leave for half an hour and then flush with a kettle of just-boiled water.

- Add half a cup of apple cider vinegar to dish-washing water to cut grease and reduce the amount of washing-up liquid needed.

Disinfecting

- Immerse a smelly sponge or dishcloth in a half and half mix of apple cider vinegar and water. Leave for two hours and then rinse with water.

- Help prevent mold discoloring bathroom tile grout by spraying tiles twice a week with water containing two tablespoons of apple cider vinegar.

- Clean a mildewed shower curtain by putting it in the washing machine along with a large bath towel. Before you start, add 4oz (100g) baking soda (sodium bicarbonate) to the washing powder in the dispenser. Then wash the load on a low-temperature setting, adding half a cup (100ml) apple cider vinegar to the fabric-softener dispenser during the rinse cycle.

Dishwasher care

- Clean a smelly dishwasher or its dispenser with a brush and soapy water, then add a cup of apple cider vinegar to the empty machine and run a cycle to remove odors.

Cleaning semi-permanent plaits, or dreadlocks

- Fill a spray bottle with 1 part apple cider vinegar and 4 parts water. Spray plaits or dreadlocks generously, leave for 10 minutes, then rinse well. This helps remove grease and hair products such as wax.

Fabric softening

- Mix 2 tablespoons of apple cider vinegar, 2 tablespoons of baking soda (sodium bicarbonate) and 4 tablespoons of water. Add to the final rinse water if washing by hand, or to the fabric-softener dispenser of a washing machine, to leave fabrics soft and static-free.

Glass cleaning

- Wipe window glass, spectacle lenses, or mirrors with a mixture of 1 part apple cider vinegar to 3 parts water, then dry with newspaper or a damp towel.

Insect repelling

- Repel dog fleas by adding half a cup of apple cider vinegar to the final rinsing water when shampooing your dog.

- Repel mosquitoes by filling a spray bottle with $\frac{1}{3}$ cup apple cider vinegar, $\frac{1}{3}$ cup witch hazel and 4 drops of citronella oil and use to spray your skin.

Limescale removing

- Soften limescale around faucets by covering overnight with a paper towel soaked in apple cider vinegar; next morning the limescale should be much easier to remove.

- Help clear limescale from a steam-iron's reservoir by filling it with apple cider vinegar. Turn on the iron, let it steam until dry, then rinse the reservoir with clean water.

Polishing

- Shine up wooden furniture by adding a few drops of apple cider vinegar to commercial polish.

- Polish wooden furniture with half and half apple cider vinegar and paraffin.

- Brighten copper and brass by applying a paste made of equal parts of salt, flour and apple cider vinegar. Let the paste dry for 10 minutes, then buff with a polishing cloth.

Rust removing

- Help remove rust by immersing small metal objects in apple cider vinegar for several hours.

Color setting

- When washing colored fabric add a cup of apple cider vinegar to help set the dye so it won't leach out and stain other fabrics.

Stain removing

- Wipe salt-stained shoes with a cup of water containing a tablespoon of apple cider vinegar.

- Clean stained stainless-steel, or copper-coated pans and bowls with a paste of salt and apple cider vinegar.

- Try removing ink, grass, coffee, tea, fruit and berry stains from fabric by soaking the stain in apple cider vinegar for an hour, then washing.

- Clean brown stains inside a tea or coffee pot by filling it with half and half apple cider vinegar and water. Leave for half an hour then rinse.

Sticky-stuff remover

- Loosen stickers or remnants of their glue by gently scrubbing the surface with apple cider vinegar.

- Use apple cider vinegar to remove the resin and hardener components of two-part epoxy glue, or even not-yet-set glue. (If any of these touch your eye or skin, irrigate the area immediately and generously with water).

- Loosen chewing gum or its stains on clothes by rubbing with apple cider vinegar before laundering.

Tights
- Make tights longer-lasting and less prone to tearing by adding a tablespoon of apple cider vinegar to the final rinse water when washing.

Washing machine care
- If your washing machine or its dispenser is smelly, clean with a brush and soapy water, then add a cup of apple cider vinegar to the empty machine and run a cycle to remove the odor.

- If your washing-machine dispenser is furred up with limescale, clean with a brush and soapy water, then add a cup of apple cider vinegar to the dispenser and run a cycle to help remove the deposits.

Weedkilling
- Kill weeds by spraying with apple cider vinegar.

Windscreen anti-icer
- Mix 3 parts apple cider vinegar with 1 part water and use this to wipe over your windscreen.

5

Recipes

From my lifelong interest in cooking, food and nutrition, I know that apples, apple juice, hard cider and apple cider vinegar can improve our health and well-being while also playing a key role in a wide variety of dishes. These tried-and-tested recipes are worth a place on anyone's table.

Apples

Apples are brilliant on their own and a good complement to many other foods. Crisp dessert apples partner well with cheese, for example, and I suggest you experiment to see which varieties of apple you prefer with which types of cheese. You might start by trying a bronzy Russet apple with a chunk of cheddar or nutty Wensleydale, a sweet Red Delicious with a piece of Stilton or Dolcelatte, or a tart Granny Smith with some crumbly goat cheese.

Sliced raw apples add excellent texture and flavor to a fruit salad and you can also add sliced, diced or shredded apple to vegetable salads – such as white, red or green cabbage salad, potato salad, beet salad, nut and celery salad and celeriac salad.

Apple slices slowly dried in the oven, cooled, then stored in an airtight container, are a great addition to picnics and lunchboxes. Apples can add a fragrant note to soups, and also form the base of many delicious kinds of chutney.

Apple sauce is traditional with roast pork, as are roasted whole apples with roast goose, and quartered pan-fried apples with pork chops. But apple sauce is also good with other meats and sausages, both hot and cold.

Sweetened stewed apple makes a quick and easy dessert that's particularly good with custard, cream, or ice cream, or as a filling for pancakes. Apple tarts, pies and crumbles are always popular and my recipe for apple and marzipan tart is ideal if you are celebrating.

- Where a recipe says "dessert" apples, it means any apples that you could enjoy eating on their own.
- Where a recipe says "cooking" apples, it means apples that are too tart to eat on their own.

Spiced Apple and Butternut Soup

**Apples lend an uncommon sweetness to this unusual and colorful soup.
You can have it hot or cold. If you prefer, use 1 tablespoon (15ml) of curry powder
instead of the four different spices.**

¼ cup (2oz, 50g) butter
2 tablespoons olive oil
1 teaspoon ground cumin
1 teaspoon ground coriander
½ teaspoon turmeric
½ teaspoon ground fenugreek
Pinch of black pepper
1 onion, peeled and sliced
2 cloves garlic, peeled
2 cups (1lb, 450g) butternut squash, peeled, seeded and cut into chunks
3 dessert apples, peeled, cored and sliced
5 cups (40fl oz, 1.1l) chicken stock
(ideally homemade by boiling a cooked chicken carcass in water with
vegetables and herbs)
½ teaspoon dried mixed herbs
1 tablespoon fresh parsley, chopped, or 4 teaspoons croutons
¾ cup (6fl oz, 180ml) sour cream – optional

- Put the butter and olive oil in a large saucepan and heat until the butter has melted. Add the cumin, coriander, turmeric, fenugreek and black pepper and fry gently for 1 minute.

- Add the onions and garlic and continue cooking for 10 minutes or until the onions are soft but not brown. Add the butternut squash, apples, chicken stock and mixed herbs. Bring to a boil and simmer for 45 minutes. Cool and blend until smooth. Stir in the sour cream and serve hot or cold. Garnish with parsley or croutons.

Sweet and Sour
Red Cabbage with Apples

This sweet-and-sour dish from central Europe complements fatty meats such as roast pork, goose and duck. It's a good idea to make double the amount as it freezes so well.

¼ cup (2oz, 50g) butter
1 large onion, peeled and finely sliced
Pinch of ground cloves
Pinch of black pepper
½ teaspoon salt
1 large red cabbage, finely shredded
2 dessert apples, peeled, cored and diced
½ cup (4oz, 100g) brown sugar
well over ¾ cup (7fl oz, 210ml) apple cider vinegar

- Melt the butter in a large saucepan. Add the onions, ground cloves, black pepper and salt. Fry gently, stirring occasionally, for about 10 minutes or until the onions are soft but not brown.

- Stir in the cabbage, apples, sugar, apple cider vinegar and enough water just to cover the cabbage. Bring to the boil, cover and simmer, stirring occasionally, for 35 minutes. Remove the lid and continue to simmer for a further 10 minutes.

Cook apples with rich meats such as pork, goose and game.

Cider-Braised Pork Sausages and Apples

Served with creamy mashed potatoes, this dish has to be the ultimate comfort food. If you prefer, you can substitute beef sausages for pork ones, in which case it's a good idea to use two large firm pears instead of the apples.

2 tablespoons olive oil
1lb (450g) large pork sausages
2 large onions, peeled and sliced
2 cloves garlic, peeled and chopped
1 cooking apple, unpeeled, cored and sliced
1 dessert apple, unpeeled, cored and sliced
1 tablespoon all-purpose flour
2 cups (15fl oz, 425ml) dry hard cider
1 tablespoon apple cider vinegar
3 bay leaves
2 teaspoons dried thyme or 2 tablespoons chopped fresh thyme
Pinch of black pepper

- Heat 1 tablespoon of oil in a frying pan, add the sausages and cook until browned all over. Remove the sausages and keep them warm.

- Add the onions to the pan and cook gently for about 10 minutes until lightly browned. Stir in the garlic and cook for a further 2 minutes.

- Heat the remaining 1 tablespoon of oil in a casserole dish, add the apple slices and cook gently, stirring occasionally, for about 3 minutes, until lightly browned on both sides.

- Add the cooked sausages, onion and garlic, then stir in the flour. Add the hard cider a little at a time, stirring to prevent lumps, then stir in the apple cider vinegar. Now add the bay leaves, thyme and pepper, cover with a lid and simmer very gently for one hour.

Baked Apples

This homely dessert is simple to prepare, fragrant and always welcome.
You can ring the changes by stuffing each apple with blackcurrant or
blackberry jam plus two teaspoonfuls of lemon juice.

4 cooking apples, cored
½ cup (3oz, 75g) golden raisins or chopped dates
⅓ cup (2oz, 50g) walnuts, chopped
just over ½ cup (4oz, 100g) brown sugar
2 tablespoons (1oz, 25g) butter, melted
just over ½ cup (5fl oz,150ml) hard cider or water

- Preheat the oven to 325ºF (170ºC, Gas Mark 3).

- Using a sharp knife, score the skin around each cored apple in a line, just above its widest part.

- Stand the apples in a buttered baking tray and well apart from each other.

- Mix the golden raisins or dates, walnuts, sugar and butter in a bowl. Use this mixture to fill the cored-out center of each apple, and pile a little extra on top. Pour the hard cider or water into the baking tray.

- Cook the apples for 30 minutes and serve with vanilla ice cream plus the scraped-out syrupy juices from the baking tray.

Apple and Marzipan Tart

This luxurious treat is perfect for high days and holidays, and if you buy ready-made pastry and marzipan it's easy to prepare.

1lb (450g) puff pastry, thawed if frozen
8oz (225g) marzipan
8oz (225g) apricot jam
4 red dessert apples, cored and thinly sliced

- Preheat the oven to 350ºF (180ºC, Gas Mark 4).

- Roll out and shape the pastry into a rectangle just larger than a non-stick baking tray. Put the pastry on to the baking tray and trim it to fit, reserving the trimmings. Prick the pastry base all over with a fork. Use a sharp knife to score a line that is 1 inch (2.5cm) in all around from the outside edge of the pastry to create a border.

- Using a rolling pin, roll out the marzipan to make a rectangular sheet the same size as the inner rectangle of pastry. Place the marzipan sheet on the pastry.

- Gently heat the apricot jam in a small saucepan until just melted.

- Lay the apple slices in overlapping rows on the marzipan and pastry base. Cut leaves, hearts or initials from the pastry trimmings, then use these as a decoration over the apples. Using a pastry brush, brush the melted apricot jam over the apple slices, pastry decorations and pastry rim.

- Cook in the oven for 20–25 minutes. Serve warm or cold, with cream.

Apple Crumble

**A good crumble with a crunchy top and soft moist inner
layer of sweetened fruit is just the ticket.**

1lb (450g) cooking apples, peeled, cored and sliced
Juice of 1 lemon
1 cup (4oz, 100g) brown sugar
just under 2 cups (7oz, 200g) all-purpose flour
1 teaspoon baking powder
¼ cup (1oz, 30g) ground almonds
just over ¼ cup (3oz, 75g) butter
½ cup (2oz, 50g) oats

- Preheat the oven to 350°F (180°C, Gas Mark 4).

- Put the apples into a buttered baking dish. Stir in both the lemon juice and
  1 tablespoon of sugar.

- Put the flour, baking powder and ground almonds into a large bowl. Cut the butter
  into little pieces and rub into the flour and almonds until the mixture resembles fine
  breadcrumbs. Mix in the remaining sugar, plus the oats.

- Level the surface of the fruit mixture and sprinkle the crumble evenly over it.

- Bake for about 30 minutes or until the crumble looks golden-brown. Serve with
  either vanilla ice cream, custard, or plain yogurt.

Apple and Ginger Chutney

The warmth of the ginger makes eating this chutney with cold meat or cheese a very special treat.

4lb (1.8kg) cooking apples, peeled, cored and chopped
2½ cups (20fl oz, 600ml) apple cider vinegar
3–4 cloves garlic, peeled and crushed
or finely chopped
3¾ cups (1½lb, 675g) sugar, dark brown
2 teaspoons ground ginger
½ teaspoon pumpkin pie spice
Pinch of cayenne pepper

- Sterilize glass preserving jars, or jam jars with screw lids, by scalding them all over with just-boiled water.

- Put the apples, half the apple cider vinegar and the garlic into a large saucepan, bring to a boil and simmer for 20 minutes or until thickened. Add the rest of the cider vinegar, and the sugar, ginger, pumpkin pie spice and cayenne pepper and cook for a further 20 minutes.

- Put the chutney in the jars and cover with waxed paper discs. When slightly cooled, cover the jars tightly with lids.

Apple Jelly

Apple jelly is easy to make from tart apples because they are so rich in pectin. Use Bramleys, other tart cooking apples, or crab apples. Eat apple jelly with bread and butter, cheese or cold meat, or add it to the saucepan when making jam from low-pectin fruits such as strawberries, raspberries, apricots, blueberries or cherries. Crab apple jelly is a traditional accompaniment for roast lamb.

4lb (1.8kg) cooking or crab apples, unpeeled, washed, cut into chunks
2 cups (1lb, 450g) sugar for each 2½ cups (1 pint, 600ml) strained juice
Juice of 1 lemon

- Sterilize glass preserving jars, or jam jars with screw lids, by scalding them all over with just-boiled water.

- Put the apple chunks into a large pan and cover with cold water. Bring to a boil and simmer for 25 minutes, or until the apples are very soft and the liquid is reduced by about a third.

- Put the apple pulp into a jelly bag, or a large strainer lined with two layers of muslin, and collect the strained liquid in a pan underneath. Leave to drip overnight. Don't push the apple through to speed collection as this would make the jelly cloudy.

- Next day, measure the liquid and put it into a large saucepan. Add the right amount of sugar, plus the lemon juice.

- Heat gently, stirring, until the sugar has dissolved. Boil rapidly for 35–40 minutes or until setting point is reached. Test for this by chilling a teaspoon in the fridge, then dipping it quickly into the jelly. Jelly at setting point will set on the back of the spoon.

- Remove the froth with a spoon.

- Put the jelly in the jars and cover with wax-paper discs. When slightly cooled, cover the jars tightly with lids.

Apple Juice

Add this ambrosial juice to fresh fruit salads, blend it with kiwifruits, mixed berries, pineapple or other fruits to make juice drinks, and delight in the character it lends to gravy or to stews of beef or pork.

Apple juice mixed with a little lemon juice and chopped apple and set with gelatin makes a wonderful jelly.

You can also use apple juice to make a range of refreshingly delicate syrups, sauces, mousses, sorbets and ice cream.

Try substituting apple juice for white wine in recipes for sauces.

Rainbow Slaw

This vibrant salad (*slaw* in dutch, as in coleslaw – meaning "cold salad") makes a great addition to a simple meal of hard-boiled eggs, cheese or cold meat or fish. Just add a crusty roll of good bread and your meal will be ready in minutes. Aim to vary the colors of the cabbage, bell pepper and grapes so as to provide an attractive range.

¾ cup (6oz,150g) cranberries, fresh, or frozen and thawed; finely chopped
2 tablespoons clear honey
2 cups (1lb, 450g) green or red cabbage, finely shredded
⅓ cup (4fl oz,120ml) apple juice
2 ribs celery, diced
1 green or red bell pepper, chopped
¾ cup (6oz,150g) black, green or red grapes, seeded and halved
⅜ cup (3fl oz, 90ml) olive or walnut oil (or a mixture)
1 tablespoon apple cider vinegar
2 teaspoons Dijon mustard
pinch of black pepper

- Mix the cranberries and honey together in a small bowl.

- Arrange all the other ingredients in a large salad bowl, add the honeyed cranberries and stir well.

Apple Juice Gravy

This makes the perfect accompaniment to pork chops or any roast pork dish as it is based on all the tasty juices that seep out during cooking.

2 tablespoons pan drippings, or pan drippings plus olive oil
1¼ cups (10fl oz, 300ml) apple juice
2 teaspoons Dijon mustard
Black pepper
¼ teaspoon cinnamon – optional

- Once you've removed the meat from the roasting pan, leave about 2 tablespoons of the meat drippings in the pan or, if there isn't enough, add some olive oil.

- Put the pan on the hob and add the apple juice, mustard, black pepper, and cinnamon if desired. Bring to a boil and simmer for 10 minutes, or until thickened.

- As a variation, you could make apple juice and sage gravy by adding 2 teaspoons of dried sage (or 2 tablespoons of chopped fresh sage) instead of the Dijon mustard and optional cinnamon .

Mulled Apple Juice

Heating and spicing transforms apple juice from a drink for the gods to a drink for partying gods.

5 cups (2 pints, 1.1l) unfiltered apple juice
2 apples, unpeeled and thinly sliced
2 oranges, unpeeled and thinly sliced
2 teaspoons pumpkin pie spice
2 bay leaves
1 cinnamon stick
1 teaspoon vanilla extract
¼ cup (2 fl oz, 60ml) dark rum, brandy or apple brandy (optional)
Few extra orange or apple slices (optional)

- Pour the apple juice into a large stainless-steel pan. Add the apples, oranges, pumpkin pie spice and bay leaves.

- Bring to a boil and simmer for 30 minutes, adding the cinnamon stick 5 minutes before the end. Add the vanilla extract.

- Strain to serve and add an optional splash of rum or brandy and decorate with a few orange or apple slices if desired.

Hard Cider

Hard cider isn't just a delicious beverage – its hints of bitterness, sweetness or sharpness, plus the unique scent bouquet and flavor of each different cider make it great to cook with too. Substituting it for stock or water adds a fragrant and unusual note to casseroled meat, poultry or vegetables.

Simmer hard cider in a saucepan until its volume has greatly reduced, then drizzle the resulting intensely flavored liquid over plain yogurt, or sweeten it as in the recipe for cider glaze, opposite.

Hard cider makes a surprisingly attractive and refreshing sorbet, and good mulled hard cider is the equal any day of good mulled wine.

Last but not least, a glass of good hard cider with a meal can be a real treat.

All the following recipes are made using dry hard cider.

Hard Cider Glaze

Brush this glaze over boiled ham before roasting, over fish, carrots or butternut squash before baking, or once they have been baked to add extra flavor. It is also good made without the honey.

2 cups (16fl oz, 480ml) hard cider
⅓ cup (3oz, 75g) butter
1 tablespoon honey

- Pour the hard cider into a pan and bring to a boil. Simmer until the hard cider has reduced to about two tablespoons.

- Remove the pan from the heat, add the butter and honey, and stir until the butter has melted.

Garlicky Chicken Casserole with Hard Cider and Apple Cider Vinegar

Ten cloves of garlic appear to be a lot but they are a vital part of this dish. The hard cider and apple cider vinegar contribute different notes to the symphony of flavors.

2 tablespoons (1oz, 25g) butter
2 tablespoons olive oil
2 onions, peeled and sliced
1 rib celery, thinly sliced
2 carrots, peeled and sliced
8 chicken thighs
½ teaspoon black pepper
1 bay leaf
1½ teaspoons dried tarragon, or 3 teaspoons fresh
10 garlic cloves, peeled
2½ cups (8oz, 225g) mushrooms, sliced
⅔ cup (4oz,100g) red lentils
2½ cups (20fl oz, 600ml) hard cider
¾ cup (6fl oz, 180ml) apple cider vinegar
Small handful fresh parsley, chopped

- Preheat the oven to 350°F (180°C , gas mark 4).

- Put the butter and olive oil into a casserole dish and heat until the butter has melted. Add the onions, celery and carrots and cook, stirring occasionally, for 10 minutes. Add the chicken thighs and black pepper and brown the chicken all over. Add the bay leaf, tarragon, garlic, mushrooms, lentils, cider, apple cider vinegar and enough water just to cover the chicken. Cover the pan and cook in the oven for 1¼ hours.

- Garnish with chopped parsley before serving.

Hard Cider Sorbet

Making sorbet with hard cider may seem an unusual idea but the result is a lovely surprise. Enjoy this refreshing sorbet on its own or with ice cream and a medley of fresh fruit, or, for a special occasion, with one or two other fruit sorbets – such as orange, apple, blackberry or plum. If you would prefer a less spicy sorbet, simply leave out the cloves, allspice berries and cardamom pod.

1¾ cups (14fl oz , 420ml) hard cider
just over ¼ cup (2oz, 50g) sugar, light brown
3 cloves
2 allspice berries
1 cardamom pod, crushed
1 cinnamon stick
3 teaspoons lemon juice

- Put the hard cider, sugar, cloves, allspice, cardamom and cinnamon into a saucepan, bring to a boil and simmer for 5 minutes.

- Cool to room temperature, strain and discard the spices. Stir in the lemon juice. Chill and freeze.

- Remove from the freezer 10 minutes before serving to allow time for the sorbet to soften a little before serving.

Mulled Hard Cider

This warming drink is a favorite in the US and the UK and makes an excellent alternative to mulled wine on a cold autumn or winter evening.

5 cups (40fl oz, 1.1l) hard cider
Juice of 1 lemon
just over ¼ cup (2oz, 50g) brown sugar
1 teaspoon pumpkin pie spice
4 cloves
1 cinnamon stick
1½ -inch (4-cm) piece of ginger root
Pinch of freshly grated nutmeg
1 red-skinned apple, cored and thinly sliced
1 green-skinned apple, cored and thinly sliced

- Put the hard cider, lemon juice, sugar and all the spices into a saucepan, bring to a boil and simmer for 15 minutes. Sieve and discard the cloves, cinnamon stick and ginger root.

- Add the apple slices, making sure to submerge them temporarily so that the surfaces facing upwards don't brown with exposure to air.

- Serve by ladling the mulled cider into glasses and sprinkling with nutmeg. Transfer one or two slices of apple into each glass too.

Apple Cider Vinegar

You can use this fragrant tawny vinegar whenever a recipe specifies malt, wine or other vinegar. It's excellent, for example, for making salad dressings and many savory sauces – including mayonnaise, mint sauce, mustard and the South American chimichurri.

Sprinkle apple cider vinegar over fried fish and chips, or over soft herring roes that have been coated with flour then fried in butter and olive oil. And improve the flavor of a gravy by adding a couple of tablespoons to the juices in the roasting pan in which you have cooked a joint of lamb.

Adding a tablespoon of apple cider vinegar to a chicken, pork or beef casserole adds interesting and attractive flavor notes.

Apple cider vinegar is a natural for pickling or marinating various vegetables and fruits.

Clam Chowder

This thick, fragrant soup originated from New England. Besides clams, it contains vegetables, milk and cream. It is popular all over the world and is a particular favorite in the US. The name "chowder" comes from the French word *chaudière* for the traditional pot in which fishermen cooked their stews of fish and shellfish.

4oz (100g) bacon, diced
2 tablespoons (1oz, 25g) butter
2 medium onions, peeled and finely chopped
2 carrots, finely sliced
2 celery ribs, chopped
1 green bell pepper, chopped
4 medium potatoes, peeled and diced
1 cup (8fl oz, 240ml) water, or water plus drained juice from the cans of clams
2½ cups (1 pint, 600ml) milk
1 cup (8fl oz, 240ml) light cream
1oz (25g) all-purpose flour
2 cups large clams, shucked (shelled) and chopped, or two 6½oz (185g) cans
minced clams
1 tablespoon apple cider vinegar
½ teaspoon Worcestershire sauce
pinch of black pepper

- Put the bacon into a large, heavy-based saucepan and fry for 5 minutes, until cooked. Transfer the bacon to a plate.

- Now melt the butter in the saucepan and stir in any bacon fat. Add the onions, carrots, celery and green bell pepper and cook for 10 minutes, stirring frequently, until lightly browned.

- Add the potatoes, water (or water and clam juice) and milk and cook, covered, for 10 minutes or until the potatoes are tender.

- Pour half the cream into a small bowl, add the flour and stir well.

- Add this cream-flour mixture to the saucepan, plus the remaining cream, clams, apple cider vinegar, Worcestershire sauce, pepper, and cooked bacon. Stir well. Bring to a light boil and simmer for 2 minutes.

Broiled Herrings
with Apple Cider Vinegar Sauce

**The tangy sauce is an ideal complement
to the richness of broiled herrings.**

8 herring fillets
2 tablespoons olive oil
Black pepper
3 tablespoons apple cider vinegar
⅓ cup (3oz, 75g) unsalted butter
1 teaspoon dried dill, or a small handful
of fresh dill, snipped

- Preheat the broiler.

- Brush the herring fillets with oil and sprinkle with black pepper. Place on a lightly oiled baking tray and broil for 2 minutes. Turn the fillets and broil for a further 2 minutes until the skin is crisped and browned.

- Meanwhile, bring the apple cider vinegar to a boil in a saucepan. Keep the cider at a simmer and cut the butter into it, whisking the mixture as it melts. Whisk in a little black pepper and the dill.

- Put the herring fillets onto warmed plates and spoon the sauce over them.

Soused Mackerel

I was brought up by the sea, and my mother frequently cooked fish. This recipe was one of our favorites. Soak up and eat some of the cooking broth with mashed potatoes as it's very rich in calcium.

4 mackerel, gutted, heads, tails and fins cut off, and washed
2 onions, peeled and sliced
2 carrots, sliced
1 rib celery, sliced
3 bay leaves
Black pepper
¾ cup (6fl oz, 180ml) apple cider vinegar
Fresh parsley or dill to decorate

- Preheat the oven to 350ºF (180ºC, gas mark 4)

- Put the mackerel, onions, carrots, celery, bay leaves and black pepper into a shallow casserole dish. Add the apple cider vinegar and enough water to cover the fish. Bake in the oven for 1 hour.

- Decorate with the parsley or dill before serving with mashed potatoes, peas and some of the juices from the dish.

Vinegared Rice Cakes with Raw Fish

Apple cider vinegar is an excellent alternative to the rice vinegar used by Japanese cooks to make sushi. Use Japanese sticky rice because it holds together well. Lay thin slices of raw fish on the rice cakes and dip the sushi into soy sauce before you eat them. Green Japanese horseradish (*wasabi*) paste is usually added to vinegared rice cakes by the cook but because it's so hot you might prefer to serve it separately. Pickled ginger makes a wonderful accompaniment. Many types of fish are suitable to eat raw, including tuna, salmon, sea bream, lemon sole, bonito and flounder. Choose top quality, extremely fresh fish. If you've never eaten raw fish before you are in for a lovely surprise.

12oz (350g) raw fish, ideally cleaned, skinned and filleted
1¼ cups (275g, 10oz) sticky short-grain rice
4 tablespoons apple cider vinegar
2 tablespoons sugar
1 tablespoon wasabi paste

- Unless the fish is already prepared, clean it, remove its scales and wash it well. Put the fish into a bowl and cover with warm water. Soak for one hour. With a thin, sharp knife, remove the skin then cut the meat from the bone.

- Chill the prepared fish for 30 minutes to make it easier to slice. Cut diagonal slices about 1½in (4cm) wide and ¼–½in (½–1½cm) thick.

- Wash the rice with several changes of water until the water runs clear, and put the rice into a heavy-based pan. Cover with 1¾ cups (¾ pint, 450ml) water, add a pinch of salt and soak for ½–1 hour. Bring to a boil and simmer, covered, for 20 minutes or until the grains are cooked but still firm. Remove the pan from the heat and leave to stand for 5–10 minutes.

- Put the apple cider vinegar and sugar into a large saucepan and heat, stirring, to dissolve the sugar. Remove from the heat, add the rice and toss it carefully to coat its grains without breaking them up.

- Take a tablespoon of the vinegared rice and form it into a firm flattened cake with your hands. Put it onto a serving dish and repeat the process with the rest of the rice to make similar cakes.

- Either put a smear of wasabi paste onto each cake and cover with a thin slice of fish, or cover the rice cake with a thin slice of fish and serve the wasabi separately.

"Pot-Roast" Lamb in Cider and Apple Cider Vinegar

Vinegary sauces are traditional with roast lamb and this luxury one-pot lamb dish is particularly easy.

Leg of lamb
2lb (900g) onions, peeled and quartered
2½ cups (20fl oz, 600ml) hard cider
½ cup (4fl oz, 120ml) apple cider vinegar
2 teaspoons dried thyme, or 4 sprigs
fresh thyme
4 garlic cloves, peeled

- Preheat the oven to 325ºF (170ºC, gas mark 3).

- Place the lamb in a large roasting pan and add the onions, hard cider, apple cider vinegar, thyme and garlic. Cover with aluminum foil tucked around the edges of the pan. Cook for 3 hours.

- Remove the lamb and onions and rest in a warm place. Strain the liquid, return this "gravy" to the pan and simmer on the hob for 10 minutes to thicken it. Carve the lamb and serve with the gravy, the onions and your choice of green vegetable.

Bone Stock

This stock, made with a cooked chicken carcass, or with ham, pork, beef, lamb or fish bones, is very rich in calcium, due to apple cider vinegar releasing calcium from the bones. Use it as the basis for a soup, or add it to casseroles or any other recipes that require stock.

Cooked stripped chicken carcass, other meat bones, or fish bones
2 carrots, peeled and finely sliced
2 onions, peeled and chopped
2 cloves garlic, peeled and crushed or chopped
¾ cup (6fl oz, 180ml) apple cider vinegar
1 teaspoon dried mixed herbs
Black pepper
½ teaspoon salt

- Put all the ingredients into a large saucepan and cover with water.

- Bring to a boil, cover and simmer for one hour, adding more water if necessary.

- Strain the stock into a bowl and use at once, or cool and freeze for another time.

Braised Pork Chops with Hard Cider, Apple Cider Vinegar, Prunes and Apples

This combination of flavors is made in heaven.
Serve with buttery mashed potatoes and steamed green cabbage.

¼ cup (2fl oz , 60ml) olive oil
4 thick pork chops, dusted with seasoned flour
8 onions, peeled and sliced
½ teaspoon ground black pepper
¼ cup (2fl oz, 60ml) apple cider vinegar
2 cooking apples, or tart dessert apples, peeled, cored and cut into
large chunks
¾ cup (6fl oz, 180ml) hard cider
4oz (100g) ready-to-eat prunes
Handful of fresh parsley, chopped

- Preheat the oven to 300ºF (150ºC, gas mark 2).

- Heat 2 tablespoons of oil in a frying pan and add the chops. Brown them on both sides and put into a baking dish or casserole. Put the remaining oil into the frying pan, add the onions and black pepper and cook for 5 minutes. Add the apple cider vinegar and apples and continue cooking, stirring frequently to prevent sticking, for a further 5 minutes. Add the cider and prunes, and cook for 5 minutes more.

- Pour the sauce from the frying pan over the chops. Cover the baking dish with foil, or the casserole with a lid, and bake for 45 minutes. Sprinkle with parsley before serving.

Oxtail Stew

This stew of succulent flavorful beef is cheap, rich in calcium and a surprise to anyone who has always thought that sirloin and fillet steak are the best cuts.

1 cup (4oz, 100g) all-purpose flour
1 stock cube
Black pepper
¼ cup (2fl oz, 60ml) olive oil
1 oxtail, cut into pieces, and with most (but not all) of the fat removed
2 onions, peeled and sliced
4 cloves garlic, peeled and crushed or chopped
4 carrots, peeled and sliced
4 ribs celery, sliced
½ cup (4fl oz,120ml) apple cider vinegar

- Preheat the oven to 300ºF (150ºC, gas mark 2).

- Mix the flour, crumbled stock cube, and pepper in a large bowl.

- Heat the oil in a large heavy-based casserole.

- Coat each piece of oxtail in the flour mixture and put into the casserole. Stir with a wooden spoon to brown the meat on all sides. Now add the onions, garlic, carrots and celery and continue stirring over the heat for 5 minutes. Add the apple cider vinegar, then enough water to cover the meat. Bring to a boil, then cover the casserole, put into the oven and cook for 2½ hours, checking every hour and adding more water if necessary.

Roast Vegetables

The magical thing about roasting vegetables is that the heat makes their sugars caramelize, which gives them an attractive, slightly nutty flavor. They are good served with roast chicken, beef and other meats. Alternatively, try serving them with goats cheese or hummus, or with brown rice or quinoa moistened with a teaspoon of soy sauce. Cut root vegetables into smaller pieces than other vegetables as they take longer to cook. Many combinations of vegetables are suitable, including leafy green ones, zucchini and – added 15 minutes before the end of the cooking time – canned red kidney beans, flageolet beans or lima beans. The following combination is an excellent one to try first.

2 onions, peeled and cut into wedges
2 carrots, peeled and sliced
2 parsnips, peeled and sliced
8 new potatoes, halved, or 2 large potatoes, cut into chunks
1 medium butternut squash, peeled, seeded and cubed
2 red bell peppers, seeded and sliced
2 cups broccoli florets
2 cups cauliflower florets
6 cloves garlic, crushed
3 tablespoons extra virgin olive oil
3 tablespoons apple cider vinegar
1 tablespoon honey
1 tablespoon dried thyme or 2 tablespoons chopped fresh thyme
1 teaspoon cumin seeds

- Preheat the oven to 375º F (190º C, gas mark 5).

- Put all the ingredients into a large roasting pan and stir well. Roast for 45 minutes or until the vegetables are cooked and lightly browned, stirring occasionally.

Braised Greens

Cooking greens this way gives a slightly sweet yet slightly tart edge to their flavor, which transforms them into an extra special treat.

1 tablespoon olive oil
2 medium onions, thinly sliced
pinch of black pepper
1½lb (675g) chopped spring greens, kale, collard greens or green cabbage,
coarsely sliced after discarding thick stems
4 tablespoons chicken stock
4 tablespoons apple cider vinegar
1 tablespoon dark brown sugar

- Heat the olive oil in a large, heavy-based saucepan, then cook the onion with the pepper for 5 minutes or until tender, stirring occasionally.

- Add the greens and cook for a further 5 minutes or until wilted, stirring frequently.

- Add the chicken stock, apple cider vinegar and sugar, cover and cook, stirring occasionally, for 15-30 minutes or until the greens are tender.

Potato Salad

Make a feast from a bowl of freshly cooked floury potatoes by adding an onion, dill and parsley salsa. This salad is particularly popular in Russia and Germany and good served hot, warm or cold.

6 medium potatoes, peeled
2 onions, peeled and chopped
3 tablespoons fresh dill, snipped
3 tablespoons fresh parsley, chopped
Pinch of black pepper
3 tablespoons walnut or olive oil,
or half and half
3 tablespoons apple cider vinegar
1 teaspoon sugar

- Put the potatoes in a saucepan of water, bring to a boil and simmer for about 20 minutes, or until soft but not falling apart.

- Drain the potatoes, cut into thick slices and put into a bowl. Add the onions, dill, parsley and black pepper.

- Put the olive oil, apple cider vinegar and sugar into a saucepan and bring to a boil, then pour this mixture over the potatoes and stir in. Serve hot or leave to cool.

Chimichurri

This popular South American barbecue sauce is arguably more of a salsa than a sauce, and it's just as delicious made with apple cider vinegar as with the red wine vinegar that's more often used in countries such as Argentina. Chimichurri can accompany any grilled, roast or barbecued meat or poultry, or you can use it as a pre-cooking marinade. It also makes a flavorsome addition to roast sweetcorn, steamed asparagus or boiled potatoes.

1 cup (8fl oz, 240ml) apple cider vinegar
1 cup (8fl oz, 240ml) olive oil
½ teaspoon cayenne pepper
½ teaspoon ground cumin
4 garlic cloves, crushed or finely chopped
1 teaspoon black pepper
1 teaspoon dried oregano, or 2 teaspoons chopped fresh oregano
2 tablespoons parsley, finely chopped
1 small onion, finely chopped
1 tomato, chopped

- Put all the ingredients into a screw-top jar, cover and shake well. Refrigerate for two hours before using.

Salad Dressing

This dressing makes lettuce leaves, raw vegetables and other salad ingredients unusually enticing. You can vary it by experimenting with different herb and spice combinations. This salad dressing can be stored in the refrigerator for up to one week.

¾ cup (6fl oz, 180ml) olive, walnut or corn oil (or a mixture of any two)
2 tablespoons apple cider vinegar
2 teaspoons Dijon mustard
1 teaspoon honey

- Put all the ingredients in a bowl and whisk well with a fork.

Variations:

Dressing for a Beet Salad:
Add 1 teaspoon of horseradish sauce.

Dressing for a Tomato Salad:
Omit the Dijon mustard from the recipe above. Add a pinch of freshly ground black pepper and 1 teaspoon of dried basil leaves (or 1 tablespoon of chopped fresh basil leaves).

Dressing for a Mushroom Salad:
Add 1 teaspoon of dried cilantro (or 1 tablespoon of chopped fresh cilantro).

Blender Mayonnaise

This homemade mayo is a real treat and is easily made in a blender. If you find the flavor of olive oil too strong, use corn or sunflower oil instead.

2 tablespoons apple cider vinegar
1 egg
2 teaspoons Dijon mustard
1 teaspoon honey
Black pepper
¾ cup (6fl oz, 180ml) olive oil, or half and half olive and walnut oils
2 tablespoons just-boiled water (optional)

- Put the apple cider vinegar, egg, mustard, honey and black pepper in the blender and blend at high speed for a few seconds until smooth. Continue blending at a lower speed and very slowly pour in the olive oil (or olive and walnut oils). If the mayonnaise is too thick, add the just-boiled water and blend for a few seconds.

Plum Sauce

This rich, sweet-and-sour plum sauce is based on the traditional Chinese sauce
that goes so well with pork spare ribs. Another idea is to serve separate dishes of
plum sauce, shredded crispy-skinned roast duck, shredded leeks, scallions finely
sliced lengthways, and three or four warmed rice pancakes per person. Diners
then fill one pancake at a time with a little of the sauce, duck, leeks and spring
onions, roll it up, and enjoy. Finger bowls of hot water scented with lemon juice
will be welcome afterwards.

If fresh plums are unavailable, use fruit from two 16oz (450g) cans of plums,
weighed after draining; then reduce the white sugar in the recipe by 2oz (50g).
Or use 4oz (100g) dried plums, rehydrated in water.

If fresh apricots are unavailable, use fruit from two 16oz (450g) cans of apricots,
weighed after draining; then reduce the white sugar in the recipe by 2oz (50g).
Or use 4oz (100g) dried apricots, rehydrated in water.

*Allow the plum sauce flavors to mature by making
and refrigerating it two weeks before it's needed.
Then warm it through before using.*

1lb (450g) fresh plums, halved and pitted
1lb (450g) fresh apricots, halved and pitted
2 cups (16fl oz, 480ml) apple cider vinegar
4 tablespoons balsamic vinegar
1 cup (8oz, 225g) brown sugar
1 cup (8oz, 225g) white sugar
Juice of 4 medium lemons
6in (15cm) length fresh ginger root, peeled and chopped
1 medium onion, peeled and finely sliced
1 hot chili pepper, seeded and chopped
2 garlic cloves, peeled and chopped
2 teaspoons salt
1 tablespoon mustard seeds, toasted
1 cinnamon stick

- Place the plums, apricots, 1 cup apple cider vinegar, ¾ cup (6fl oz, 180ml) water and balsamic vinegar into a large saucepan. Bring to a boil then simmer for 15 minutes.

- Put the remaining apple cider vinegar with the brown sugar, white sugar and lemon juice into another saucepan. Bring to a boil, simmer for 10 minutes then cool for 5 minutes before adding to the vinegared fruit in the first pan.

- Stir in the ginger, onion, chili, garlic, salt, mustard and cinnamon. Bring to a boil and simmer for 45 minutes. Discard the cinnamon stick.

- Pour the mixture into a blender and blend until smooth. Return to the saucepan, bring to a boil and simmer for 15 minutes or until thick.

- Ladle the sauce into a sterilized preserving jar. Cover loosely and cool for three hours. Cap the jar and refrigerate.

Cucumber Pickle

**This is brilliant with cold savory food and a sure-fire hit whether served for
solo repasts, family meals or festive gatherings.**

2lb (900g) cucumber, peeled and finely sliced
3 cups (1lb, 450g) onions, peeled and finely sliced
2 tablespoons salt
just under 2 cups (15fl oz, 450ml)
apple cider vinegar
3½ cups (12oz, 350g) brown sugar
½ teaspoon turmeric
½ teaspoon ground cloves
4 teaspoons mustard seed
4 teaspoons celery seed (optional, but worthwhile if you can get it)

- Sterilize glass preserving jars, or jam jars with screw-on lids, by scalding them all
  over with just-boiled water.

- Put the cucumber, onions and salt into a bowl, mix well and leave for 3 hours. Rinse
  well under cold running water and strain.

- Put the cucumber and onions into a large saucepan, add the apple cider vinegar
  and bring to a boil. Simmer gently for 20 minutes. Add the sugar, turmeric, cloves,
  mustard seed and celery seed, if using, and stir until the sugar has dissolved. Bring
  to a boil then simmer for 2 minutes.

- Remove the cucumber and onions with a slotted spoon and put into the warm glass
  jars. Simmer the remaining syrup for 15 minutes, then pour it over the cucumber
  and onions. Cover the jars tightly.

Beets and Horseradish Relish
(Cwikla or *Red Chrain)*

This colorful accompaniment for fish, meat or cheese originated in Eastern
Europe and in Russia. Once you've tried it, the odds are that you'll become a fan.

1lb (450g) raw beets, washed
2 tablespoons horseradish sauce
1 tablespoon wholegrain mustard
¼ cup (2fl oz, 60ml) apple cider vinegar
1 tablespoon sugar
Plenty of black pepper

- Put the whole beets into a saucepan of water, bring to a boil and leave for at least
  30 minutes or until tender when tested with a knife. Leave to cool, then rub off the
  skins and grate the beets into a bowl.

- Stir the horseradish sauce, mustard, apple cider vinegar, sugar and pepper into the
  grated beets.

Marinated Pears

Tickle your taste buds by eating these sweet-and-sour pears with cold meat, sausage or cheese.

2lb (1kg) hard pears, peeled, cored and quartered
2½ cups (1 pint, 550ml) apple cider vinegar
2¼ cups (1lb, 450g) sugar
1 clove
Pinch of cinnamon
1 bay leaf
1 teaspoon black peppercorns
Pinch of salt

- Sterilize glass preserving jars, or jam jars with screw-on lids, by scalding them all over with just-boiled water.

- Cover the pears with water in a pan, bring to a boil and simmer for 15 minutes or until slightly soft. Drain and cool in the sieve under cold running water.

- Put the apple cider vinegar, 2½ cups (1 pint, 550ml) water, sugar, clove, cinnamon, bay leaf, peppercorns and salt into the pan, bring to a boil and simmer for five minutes. Gently add the pears, bring to a boil again, then leave to cool. Put the pears into sterilized glass jars, fill with the liquid, then screw on the lids.

Apricot Relish

A spoonful of apricot relish makes a wonderful addition to roast lamb or pork, or potatoes roasted in their skins with garlic and thyme. It is equally good served hot or cold.

1 cup (5oz,150g) dried apricots
¾ cup (6fl oz,180ml) apple juice
¾ cup (6fl oz ,180ml) apple cider vinegar
2oz (50g) soft brown sugar
1 teaspoon dried cinnamon
1 teaspoon freshly grated nutmeg
Pinch of ground black pepper
½oz (12.5g) butter

- Put the apricots, apple juice, apple cider vinegar and sugar in a saucepan. Bring to a boil and simmer gently for 10 minutes. Leave to cool a little.

- Put into a blender, add the spices and butter and blend until smooth.

- If you prefer to serve the relish hot, return it to the pan and heat through gently.

Fondant Zucchini

**The basis of this recipe was kindly given to me by the owner of the amazing
Blairs Cove House restaurant in Durrus, County Cork, Ireland, after I'd waxed
lyrical about it during an evening there. I've adapted the recipe to use apple cider
vinegar instead of malt vinegar, and I promise it's worth every last one of the four
days it takes to make. Use to accompany cold meats, herring or cheese.**

*6 zucchini, or 1 medium vegetable marrow, peeled, seeded and
cut into chunks*
2 large onions, peeled and sliced into rings
3 tablespoons salt
2 cups (16fl oz, 480ml) apple cider vinegar
2½ cups (18oz, 500g) sugar
1 tablespoon curry powder
1 teaspoon black peppercorns

Day 1
- Put the zucchini and onions into a bowl and stir in the salt.

Day 2
- Drain the zucchini and onions in a large colander and rinse well under cold running water.

- Put the apple cider vinegar, 2½ cups (20fl oz, 600ml) water, 14oz (400g) sugar, curry powder and black peppercorns into a large saucepan. Bring to a boil, add the zucchini and simmer for 5 minutes.

- Transfer to a large bowl and cool at room temperature.

Day 3
- Stir the remaining sugar into the mixture in the bowl.

Day 4
- Put the mixture into a large saucepan, bring to a boil, and boil for 5 minutes.

- Sterilize glass preserving jars, or jam jars with screw-on lids, by scalding them all over with just-boiled water.

- Put the mixture into the jars. Cool slightly and cover tightly with lids.

Tips for Cooks

Apples and apple cider vinegar can be a great help to cooks.

Apples

Cake
Keep a cake fresher longer by putting it in an airtight container along with a halved apple.

Over-salted soup or casserole
Add a few peeled apple chunks to soak up excess salt, then remove after 10–15 minutes.

Tomatoes
Speed up the ripening of unripe tomatoes by putting them in a paper bag with one ripe apple for each three tomatoes, for a few days. The apples release ethylene gas which speeds ripening of the tomatoes.

Apple Cider Vinegar

Beans
Discourage flatulence by adding a tablespoon of apple cider vinegar to the water when soaking dried beans.

Cheese
Help prevent stored cheese hardening by wrapping it in muslin or cheesecloth soaked in apple cider vinegar.

Eggs

Poaching: put 2 teaspoons of apple cider vinegar in the water to help egg whites stay better formed.

Hard-boiling: put 1 or 2 tablespoons of apple cider vinegar in the water to make eggs easier to shell.

Boiling: put 2 tablespoons of apple cider vinegar in the water to help prevent shells cracking.

Jellies or jellied savory dishes

Add a teaspoon of apple cider vinegar to the warm liquid to help gelatin set.

Meat and fish

When marinating, braising, poaching or boiling meat, or poaching fish, add half a cup of apple cider vinegar to each cup of liquid to make the meat or fish more tender and to draw calcium from its bones.

Meringue

Add a teaspoon of apple cider vinegar to every 2 egg whites and leave to stand for 30 seconds before whipping. This increases their stiffness and makes meringues brilliantly white.

Pancakes

If you'd like to use buttermilk but you haven't any, add a tablespoon of apple cider vinegar to a cup of milk and leave it for five minutes before using.

Pastry

Instead of adding water to the flour-butter mixture, add flavor by adding apple cider vinegar, or half and half of apple cider vinegar and water.

Rice or pasta

Put a teaspoon of apple cider vinegar into the cooking water and you'll find the cooked rice or pasta is less sticky.

Salads, vegetables and fruit

Washing with an apple cider vinegar solution may help remove certain pesticides and potentially harmful bacteria. To make the solution, mix 1 part apple cider vinegar to 9 parts water, immerse the produce and let it soak for 5 minutes, then rinse well.

Soups, gravy or a savory sauce

Add 2 tablespoons of apple cider vinegar to improve the flavor.

Stock made with a chicken carcass or other bones

Add a tablespoon of apple cider vinegar to the water to enrich the stock with calcium from the bones.

Vegetables

When boiling or steaming vegetables, add a splash of apple cider vinegar to the water while cooking to help the vegetables retain their color.

6

Make Your Own Apple Juice, Hard Cider & Apple Cider Vinegar

Home-pressed fresh apple juice and home-brewed hard cider and apple cider vinegar have distinctive flavors and give you the satisfaction of knowing exactly what they contain. Cloudy apple juice and hard cider contain tiny fragments of suspended apple pulp and because of this they are richer in pectin, phenolic acids and certain other health-promoting phytochemicals than are the clear juice and hard cider varieties, which may have water added.

Very, very important from the outset: when making apple juice, hard cider or apple cider vinegar, always choose containers made from food-grade stainless steel or plastic, or glass. Never allow cut-up or crushed apples, apple juice, or hard cider to come into contact with any containers, utensils or equipment that have any iron, copper or lead content.

Cleanliness is crucial at all stages of each process. To avoid spoilage of juice, hard cider or apple cider vinegar by unwanted microorganisms, you must wash and scrub all containers, utensils and equipment extremely thoroughly, using

BEST JUICING APPLES

Although thousands of apple varieties are known and grown, about 20 dominate the market and are readily available in shops. Of these the most popular apples for juicing are Golden Delicious, Granny Smith, Idared, McIntosh, Rome and York. These tend to be full of juice and very sweet in flavor. Experiment by mixing with other varieties to achieve the flavor that refreshes you best.

hot water and no soap. Then rinse everything well with just-boiled water. Your kitchen and your hands must also be very clean. Some people recommend sanitizing all equipment and bottles with a sulfite solution (for example, made by adding Campden tablets to water); they may also add this to their fermenting cider to kill unwanted bacteria. However, quite a few people are sensitive to sulfites. Also, the amounts of sulfite used often do not completely sterilize and may result in leaving some unwanted bacteria untouched.

Apple Juice

The most important decision you'll make before you start the juice-making process is that of selecting which variety or varieties of apple to use, as this determines the proportions of sugar and acid that will be present in the juice. Dessert apples tend to be sweet. Cooking apples tend to be more acidic, or tart. Blending the two types – for example, two-thirds sweet apples to one-third tart apples – enables you to adjust the sweetness and acidity of the resulting juice. Cider apples are extremely sour and bitter and therefore are less suitable for making apple juice for drinking, but added in a small quantity they can boost the overall flavor. All apples must be thoroughly clean, worm free and have no trace of mold or decay.

To make cloudy apple juice:

Step 1

Select the apples. They should be newly picked, firm and shiny and definitely unbruised, undamaged and rot-free. They can be of one variety, or several – three varieties is often recommended, for example. Strongly flavored dessert apples give a good flavor, especially if you add crab apples for a hint of bitterness from their high tannin content. Bittersweet and bittersharp varieties of cider apples have relatively high tannin levels too. The sweeter apples are, the more alcoholic their juice becomes. Fall-gathered apples tend to be sweeter than summer ones. As a very rough guide, about 20lb (9kg) of apples yields approximately 1 gallon (4 liters) of juice.

Step 2
Wash the apples thoroughly in cold water.

Step 3
Cut up or crush the apples. Either cut the apples into tiny pieces that are smaller than peas, or crush (pulp, grind or mince) them using a food processor, blender or juicer with a strong motor, or a fruit mill or crusher (from a wine-makers' supplier). They could even be pulped by fitting a pulping attachment (a blade called a Pulpmaster) to an electric drill, or crushed (very carefully) with a hammer. Crushed apple pulp is called "pomace" and rather than throw it away you can add any excess to cut-up whole apples when making an apple pie.

Step 4
Press the crushed or cut-up apples. Use an apple press (bought or hired from a wine-makers' supplier) to extract the juice from the milled apples by cold-pressing. Newly pressed apple juice goes brown within a few minutes, and it is this "tanning" that is largely responsible for the final color of the juice.

Either drink the apple juice at once, or keep it in the fridge for 7–14 days (any longer and it will begin to ferment). The cloudiness may settle as sediment at the bottom of the bottle so, if necessary, stir or shake the juice before you drink it.

You can help prevent the natural browning of apple juice (caused by the oxidation of its tannins) by adding ¼ teaspoon (1 gram) of powdered vitamin C (ascorbic acid, from a wine-makers' supplier) to each 4 pints (2 liters) of freshly pressed juice.

Some people make large quantities of juice and preserve it by freezing, pasteurizing (which destroys some of its vitamin C content), or chemical treatment. Preservation prevents microorganisms "spoiling" (fermenting) the juice and thereby creating gases which build up pressure in capped bottles that could make them explode. Freezing is the best method for preserving homemade apple juice.

Freezing

To freeze apple juice, pour it into plastic containers, filling them not quite full (to allow for expansion), then cover and freeze without delay. Frozen juice keeps well for at least a year. Shake thawed juice before drinking it, as its valuable cloudiness tends to settle as sediment. Unless you pasteurized the juice (see below) before freezing it, keep it in the fridge and be aware that it will not keep for long as the enzymes and yeasts will become active again after thawing.

Pasteurizing

To pasteurize apple juice, sterilize glass Kilner jars, or glass bottles that can be capped, by pouring boiling water into them and all over the jar lids and bottle caps. Now fill them with the juice, leaving an inch free at the top. Put the open jars or bottles in a large pan, fill the pan with water to a level of about 2in (5cm) below the open top of the lowest jar or bottle, and heat the water to 167°F (85°C), checking with a thermometer. Simmer at this temperature for 10 minutes

BEST CIDER APPLE VARIETIES

Hard cider was very common in America in colonial times and drunk to quench the thirst as a safer alternative to untreated water. The apple varieties used by settlers are still popular with cider-makers today and include: Baldwin, Harrison, Roxbury Russet, Vista Bella and Winesap. The flavor of cider apples varies with their variety, being bittersweet, bittersharp, sweet or sharp, so a cider's flavor depends on the blend of apples.

in order to destroy the apples' natural yeasts. Remove the bottles, put them on a wooden or plastic board, and close or cap them when the bottles are cool enough to handle.

Preserving

To preserve the juice chemically, add the food preservative potassium sorbate (available as a wine stabilizer from wine-makers' suppliers; use as directed on the packet). This does not inhibit enzymes or all bacteria though, so the flavor and color of the juice may deteriorate after a few days, and the cloudiness may eventually settle out.

Commercial apple juice may be cloudy (unfiltered) or clear (filtered). To clarify cloudy juice, it is first left to settle overnight; the clear juice is then siphoned from the sediment (racked) then filtered. Commercial juice is usually pasteurized and may also be treated in other ways. The label may give more information.

Hard Cider

Alcoholic cider is made from apples by allowing or encouraging fermentation of natural apple sugars (and, perhaps, of added sugar) to alcohol. This is known as "yeast" or "alcoholic" fermentation.

Traditionally made cloudy hard cider is much richer in health-promoting proanthocyanidin plant pigments than are many red wines – and certainly more so than most commercial hard ciders.

The two main reasons for making hard cider at home are that it's both cheap and easy to produce delicious cloudy hard cider.

Another reason is that you can produce hard cider free from added sulfite, whereas most commercial hard ciders have had sulfite (or bisulfite or metabisulfite) added to help prevent spoiling. The concern is that perhaps as many as 1 percent of us are sensitive to sulfites, and some people are seriously so. Sulfite sensitivity is not a true food allergy, but its symptoms are similar. If you are buying hard cider and want to avoid added sulfite, read the label.

Making hard cider

This is how to make cloudy hard cider at home:

Step 1
Make a batch of cloudy apple juice, including some cider apples (*see* the box opposite), as described on the previous pages.

Next you can take one of two approaches. Either:

Step 2a

This method takes a slightly more scientific approach to treat the juice and so help ensure a successful brew by measuring sugar, acidity and pH. If necessary, blend in some extra juice from apples with more sugar or acidity or a lower pH.

Sugar level: In a good summer, the sugar level of apple juice may be as high as 17 percent; in a cool wet one it may be less than 10 percent. You can estimate the sugar content – and therefore the likely alcohol content of the finished hard cider – with a hydrometer (a gadget available from wine-makers' suppliers) which measures the specific gravity ("heaviness") of the juice. For example, a specific gravity of 1.070 suggests a sugar content of 15 percent and an eventual alcohol content of 8.5 percent. And a specific gravity of 1.045 suggests a sugar content of 10 percent and an eventual alcohol content of 6 percent. If your juice has a specific gravity of less than 1.045 and you have no sweeter juice to blend with it, you may want to add sugar; if you don't, the alcohol content of the finished hard cider may not be high enough to prevent it spoiling. Add 1 tablespoon of sugar to each 2 pints or quart of juice and stir well. Retest with the hydrometer and repeat if necessary.

Acidity and pH: An apple's acidity is determined more by its variety than by the climate. It is useful to know the acid content and pH (acid-alkaline balance) of the juice. Ideally, it should have a malic acid content of 0.3–0.5 percent. If it contains less, the pH will be too high and fermentation will be susceptible to bacterial infection. If it contains more, the pH will be too low and the finished hard cider will taste too sharp. One option is to measure the acidity with a titration kit, and measure the pH with a pH meter (both available from wine-makers' suppliers). Aim for a pH of 3.2–3.8. Many bittersweet cider apples have a high pH, so need blending with more acidic fruit.

Or: *Step 2b*
Just taste the juice! If, like many home cider-makers, you prefer not to bother with measuring sugar, acidity and pH, and be content to accept a slightly higher risk of failure, simply taste the juice:

- If it tastes insipid, and you have no other juice available for blending, add malic acid in doses of ¼ teaspoon per 2 pints and keep tasting it until you think its flavor has improved.

- If it tastes too acidic, and you cannot blend the sharpness out with other juice, add malic acid (in the recommended dose) to encourage malolactic fermentation, or alternatively, add calcium carbonate to neutralize the acid (in a dose of ¼ teaspoon per 2 pints and repeated as necessary). Malic acid and calcium carbonate (precipitated chalk) are available from any wine-makers' suppliers.

Step 3
Strain the blended juice through a coarse sieve, pour it into a glass demijohn or carboy (from a wine-makers' supplier), and put this in a cool dark place, at around 50°F (10°C) or lower.

Step 4
Allow the juice to ferment. Letting wild yeasts and bacteria ferment the juice can work well but may produce cider with vinegary or other unwanted flavors. Alternatively, you can speed up fermentation and make the result more reliable by adding wine yeast plus sugar syrup, though this may produce a somewhat bland tasting cider.

If you decide to add yeast:
- Add wine yeast (not brewer's or baker's yeast!) from a wine-makers' supplier, according to the instructions on the packet.
- Add a teaspoon of sugar or honey per 2 pints (1 liter) of cider.
- Consider encouraging efficient fermentation by adding yeast nutrients: add 0.2mg of thiamine and up to 300mg of ammonium salt per liter.

Fermentation usually begins within 48 hours. Early on, there is considerable bubbling caused by the release of carbon dioxide as the yeasts multiply and break down the sugar. This usually continues for about three days. When it subsides, fit an airlock or fermentation tap (from a wine-makers' supplier) so carbon dioxide can get out but air can't get in. You should see bubbles escaping through the airlock within two to three days; this will continue for one to three months, until all the sugar has been fermented into alcohol.

ENHANCING FLAVOR DURING FERMENTATION

Wine yeast produces alcohol by acting on natural sugars in the fermenting juice. But adding 2 teaspoons of sugar or honey makes the hard cider stronger by letting wine yeast produce extra alcohol. The flavor components of honey can add subtle variations in flavor to the finished cider. They vary with the source of nectar from which the honey is made. So you could experiment by using a different honey for each batch of hard cider.

When you don't see any more bubbles go through the airlock during five minutes of watching, remove the airlock and use a clean plastic tube to siphon the clear cider into another container, leaving the sediment behind. This is called "racking" the cider. Rinse the original container well, pour the cider back in and replace the airlock. If the airlock's "thimble" goes back up within 24 hrs, leave the cider for a week then rack it again. Repeat the racking three times if necessary.

Step 5
Bottle the cider once the airlock thimble no longer comes back up within 24 hours of racking and you are absolutely sure that no more fermentation is taking place. Ideally, store bottled cider for three to six months before drinking. This improves its flavor by allowing continuing malolactic fermentation, in which malic acid is converted to lactic acid.

Apple Cider Vinegar

Apple cider vinegar results from the fermentation by acetobacter bacteria of the sugar in hard cider to acetic acid. It's easier to make homemade apple cider vinegar than to make hard cider, because yeast is the only thing you'll need to add. The flavor of homemade apple cider vinegar is often more delicate and complex than that of the commercially produced variety, and because it has not been pasteurized, its flavor may continue developing over several years.

Most commercial producers convert cider to apple cider vinegar within a few hours by using a large fermenter with forced aeration and added acetobacter. Domestic apple cider vinegar makers cannot buy these bacteria and most commercially produced vinegars do not work as a starter to set off and encourage fermentation, since they have been pasteurized and so contain no acetobacter. Wild acetobacter bacteria eventually find their way into hard cider, but adding an active starter such as brewer's yeast aids fermentation. Use wooden, glass or food-grade stainless steel containers when making or storing apple cider vinegar. Do not use other metal, plastic, or glazed ceramic containers.

Making apple cider vinegar

Step 1

Make apple juice as described on pages 166–170, noting that the sweeter the apples, the stronger the apple cider vinegar will be.

Step 2

Consider adding brewer's yeast (according to the instructions on the packet); while not essential, this hastens alcoholic fermentation.

Step 3

Put the juice into an open container, filled only three-quarters full, to ensure easy entry of acetobacter bacteria, cover with a muslin cloth tied around the rim of the container to exclude insects, and leave in a warm dark place. It should start bubbling within a few days.

Step 4

Aerate the mixture each day by stirring vigorously, and keep warm at around 65–86ºF (18–30ºC). Gradually a whitish gel-like raft of "vinegar mother" will form. This contains acetobacter bacteria plus the cellulose they make to keep them floating, since they need plenty of air.

Step 5

Consider speeding up fermentation by adding some previously made unpasteurized and preservative-free apple cider vinegar. Not only do acetobacter thrive in a more acidic environment, but the added vinegar will probably supply some live acetobacter from remaining traces of vinegar mother, and these will act as a starter. Add about half a cup or 4fl oz (just under 100ml) to approximately every 4 pints (i liter) of fermenting liquid.

Step 6

Leave the apple cider vinegar undisturbed for four weeks if you have added yeast, eight weeks if not, then taste it. If you think it is vinegary enough, siphon the rest into sterilized bottles, filling them to the top and capping them. If not vinegary enough, leave it for as long as it takes, tasting every week. It is better not to filter apple cider vinegar. And there is no need to pasteurize small quantities of homemade apple cider vinegar. Traces of vinegar mother may show as slight cloudiness or as gelatinous particles.

Useful Addresses & Websites

Here is a selection of the many companies and organizations concerned with apples, apple products, and apple cider and vinegar making.

United States
Apple Products Research & Education Council
Tel: +1 404 2523663.
www.appleproducts.org
Provides information on apples and apple products.

Leeners
www.leeners.com
Tel: +1 800 543 3697
Supplies equipment for making apple juice and cider.

US Apple Association
www.usapple.org
Promotes apples and apple products and provides information to consumers, educators, the media and industry.

USDA (United States Department of Agriculture)
www.usda.gov
Provides information on apples, apple juice, cider and cider vinegar.

These websites provide state-specific and other information about apples:

www.bestapples.com (Washington)

www.calapple.org

www.michiganapples.com

www.nyapplecountry.com

United Kingdom

Aspalls

Tel: 0044 1728 860510

www.aspall.co.uk

Makes preservative-free, unpasteurized apple juice, "cyder" and "cyder" vinegar.

Boots the Chemist

Larger branches carry a wide range of wine-making chemicals and sundries.

Bramley Apple Information Service

Tel: 0044 20 70528951

www.bramleyapples.co.uk

Offers on-line recipes and information about apples, as well as recipes in a booklet and a quarterly e-letter.

Brogdale Collections

Tel: 0044 1795 536250

www.brogdalecollections.co.uk

Houses the UK's National Fruit Collection with over 2,300 varieties of apple tree –

the largest collection in the world. The collection is open to the public, along with exhibitions, demonstrations, talks, day schools and workshops.

Common Ground
Tel: 0044 1747 850820
www.commonground.org.uk
Publishes information about apples and co-ordinates National Apple Day (in October).

English Apples & Pears Limited
Tel: 0044 1732529781
www.englishapplesandpears.co.uk
Represents the industry and promotes English apples and pears.

H P Bulmer Limited
Tel: 0044 1432 352000
www.bulmer.com
The world's largest maker of cider, including the Strongbow, Woodpecker and Scrumpy Jack brands.

National Association of Cider Makers
Tel: 0044 117 3178135
www.cideruk.com
Represents cider producers and promotes the cider industry in the UK. Has an on-line biannual newsletter, "Cider Matters."

www.orangepippin.com
This website for apple enthusiasts and orchardists describes the flavors of apples and the origins of different varieties. It also lists its own top 10 apples and the top 10 apples chosen by consumers.

Somerset Cider Vinegar Co
Tel: 0044 1278 723292
www.somersetcidervinegarco.co.uk
Uses cider apples to produce cider vinegar that is aged for 2 years.

The Brew Shop
Tel: 0044 161 480 4880
www.thebrewshop.com
Cider-makers' supplies, including apple presses for sale or hire.

The Campaign for Real Ale (CAMRA)
Tel: 0044 1727 867 201
www.camra.org.uk
CAMRA has a subgroup, The Apple and Pear Produce Liason Executive (APPLE), which aims to protect traditional English varieties of cider and perry; it also publishes the *Good Cider Guide* which lists pubs in Britain that offer real cider.

Wineworks
Tel: 0044 1246 279382
www.wineworks.co.uk/product/pulpmaster/
Cider-makers' supplies, including a Pulpmaster (a pulping blade that you fit to an electric drill) and a manual apple crusher.

Vigo Limited
Tel: 0044 1404 892101
www.vigopresses.co.uk
Cider-makers' supplies, including apple presses and crushers.

Australia

Apple and Pear Australia Ltd
www.apal.org.au
Represents commercial apple growers in Australia; its website provides recipes and health and other information about apples.

Have an Aussie Apple
www.haveanaussieapple.com
This website is funded by Horticulture Australia Ltd and offers consumers apple-related recipes and health and other information.

New Zealand

Horticulture New Zealand
www.hortnz.co.nz
Represents commercial growers.

Pipfruit New Zealand
www.pipfruitnz.co.nz
Tel: +64 6 873 7080
Represents the New Zealand pipfruit industry; its website provides information, news and links to other apple-related websites.

Other

The World Apple and Pear Association
www.wapa-association.org
Represents major apple- and pear-producing countries globally and has a news section.

Index

Notes